DIFFERENTIAL TOPOLOGY

An Introduction

DAVID B. GAULD

Department of Mathematics
University of Auckland
Auckland, New Zealand

DOVER PUBLICATIONS, INC.
Mineola, New York

Copyright

Copyright © 1982, 2006 by David B. Gauld
All rights reserved.

Bibliographical Note

This Dover edition, first published in 2006, is an unabridged, slightly corrected republication of the work originally published by Marcel Dekker, Inc., New York, in 1982. Some of the figures have been updated for this edition.

Library of Congress Cataloging-in-Publication Data

Gauld, David B.
   Differential topology : an introduction / David B. Gauld.
     p. cm.
   Originally published: New York : M. Dekker, 1982.
   Includes bibliographical references and index.
   ISBN 0-486-45021-X (pbk.)
     1. Differential topology. I. Title.

QA613.6.G38 2006
514'.72—dc22

2005056933

Manufactured in the United States of America
Dover Publications, Inc., 31 East 2nd Street, Mineola, N.Y. 11501

PREFACE

The aim of this book is to present a classification of surfaces from the viewpoint of differential topology. Along the way the reader is introduced to topological spaces then proceeds to a study of differentiable manifolds. Morse theory and surgery, which occupied differential topologists particularly in the 1950s and 1960s, are studied at a fairly general level before attention is restricted to the effect of surgery on a surface. This surgery then leads to our classification.

The audience to which this book is aimed consists of senior undergraduates, the main prerequisite being a course in multi-variable calculus having its foundations in linear algebra. A rigorous course in introductory analysis is not assumed, the necessary topology being given in the first three chapters. The essential calculus is then summarized in Chap. 4. Thereafter the book is basically self-contained, although there is reference to a few deeper results which are not proved here. These results are collected together in Appendix 1, where there are either proofs of them or indications of their proofs together with references where complete proofs may be found.

For three years since 1977, versions of this book have been used in a third-year course based on the Personalized System of Instruction at the University of Auckland. After an introductory lecture, students were issued a copy of the first chapter. Thereafter they were allowed to proceed to a new chapter only after attaining a mastery level score in a test on the previous chapter.

Successful completion of a chapter resulted in the student receiving four points toward the final grade in the course. The remaining 40 points came from a final examination, several versions of which were offered at various times. A student could complete the whole book in a 15-week semester. In other years, the book has been used as a text in a regular lecture course.

The graph below shows the way in which chapters are interrelated.

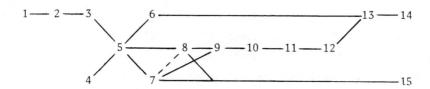

The content of this book can be understood at several different levels, ranging from a slight modification where necessary of the reader's prior level of intuition, to a complete grasp of all concepts and proofs. Full details are given in most cases even though it is not expected that all readers will reach this level on a first journey. Students should, at the least, read everything and try to relate it to their intuitive ideas of the various topics. This is particularly the case with Chaps 1, 6, 8, and 10. An important way of relating these concepts to intuition is to draw pictures. No apology is made for the large number of figures in the text, although these run a distant second place to one's own well-conceived pictures.

Thanks must be extended to the students of the Mathematics 350 class at the University of Auckland in 1977 for permitting themselves to be used as guinea pigs for the first tentative version of this text as well as the method of instruction. Particular thanks are due to the pioneers, Jennie, Michael, Richard, and Brian, who at various times were the first to attack a particular chapter and so induced me to rewrite better versions for the rest of the class. Thanks go to Mrs. Gladys Hubbard and Mrs. Helen Sparkes for typing early drafts, and to Miss Eve Malbon for her excellent typing of the final camera-ready copy.

David B. Gauld

CONTENTS

Preface		v
1.	What Is Topology?	1
2.	Topological Spaces	15
3.	Some Topological Properties	27
4.	Some Advanced Calculus	39
5.	Differentiable Manifolds	53
6.	Orientability	65
7.	Submanifolds and an Embedding Theorem	81
8.	Tangent Spaces	93
9.	Critical Points Again	107
10.	Vector Fields and Integral Curves	123
11.	Surgery	139
12.	The Trace of a Surgery	155
13.	Surgery on a Surface	169
14.	Classification of Orientable Surfaces	189
15.	Whitney's Embedding Theorem	207
Appendix A.	The Unproved Theorems	217
Appendix B.	Further Topics	225
Notation		235
Bibliography		237
Index		239

DIFFERENTIAL TOPOLOGY

An Introduction

WHAT IS TOPOLOGY?

I. A mathematician confided
   That a Möbius band is one-sided,
   And you get quite a laugh
   When you cut one in half,
   For it stays in one piece when divided.

II. A mathematician named Klein
    Thought the Möbius band divine.
    He said, "If you glue
    The edges of two,
    You get a bottle like mine!"

III. A topologist can remove his shirt while wearing his jacket.

IV. Tie the wrists of two people together with two lengths of string as in Fig. 1. It is possible for them to disengage themselves without slipping the string off their wrists, breaking the string or untying any knot.

V. A topologist cannot distinguish a coffee cup from a (ring) doughnut.

VI. Topology is the study of topological invariants.

I to V, although whimsical, give a vague idea of the kind of problem met by topologists. I illustrates the problem of orientability: the one- or two-sidedness of an object. We will meet this problem in Chap. 6. II illustrates a technique of constructing new objects

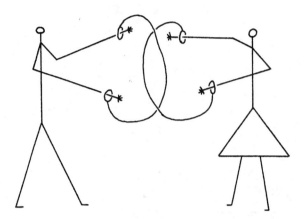

FIGURE 1

from old found in many different areas, topology one of them. We will meet this technique in Chap. 11. The party tricks III and IV, together with V, illustrate the intuitive feeling that we can somehow deform one configuration (the shirted topologist, the shackled couple or the coffee cup) without tearing but by stretching, shrinking and twisting to end up with the other configuration. Of course in the case of III and IV certain objects would resist much stretching and shrinking! In each case the deformation of one configuration into the other is an example of a homeomorphism, one of the basic concepts of topology. VI is a more formal definition of topology, which will now be explained.

Let

$$\mathbb{R}^n = \{(x_1,\ldots,x_n) \mid x_i \in \mathbb{R}\}$$

where $\mathbb{R}$ is the set of real numbers. We identify $\mathbb{R}^1$ and $\mathbb{R}$, and represent it geometrically by a line in the usual way. Similarly, $\mathbb{R}^2$ may be represented geometrically by a plane, x_1 and x_2 being,

1. What Is Topology?

respectively, the x and y coordinates of a point, and $\mathbb{R}^3$ may be represented geometrically by three-dimensional space. In general, $\mathbb{R}^n$ is n-dimensional space, though we three-dimensional beings find a geometrical visualization of $\mathbb{R}^n$ (n > 3) rather difficult. Nevertheless, topologists spend much time drawing pictures to inspire them. Although these pictures are two-dimensional, they often exhibit the kinds of problems to be overcome. Get into the habit of drawing pictures.

For $x = (x_i)$, $y = (y_i) \in \mathbb{R}^n$, let $|x-y|$ denote the usual *pythagorean distance* from x to y, i.e.,

$$|x - y| = \sqrt{\sum_{i=1}^{n} (x_i - y_i)^2}$$

With this distance, $\mathbb{R}^n$ is often called *euclidean space*. Some people require euclidean space to have more structure (e.g., the vector space structure) but it does not really matter here.

Suppose $x \in \mathbb{R}^n$, $A \subset \mathbb{R}^n$. Say that x *is near* A and write x ν A iff for all r > 0, there exists a ∈ A with $|x-a| < r$. Use the notation x ̸ν A to mean that x ν A is false, i.e., x is not near A (Fig. 2).

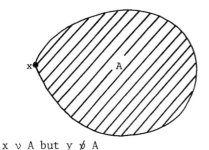

x ν A but y ̸ν A

FIGURE 2

Notice that the relation ν between points of $\mathbb{R}^n$ and subsets of $\mathbb{R}^n$ satisfies the following basic properties:

Near 1. $x \nu A \Rightarrow A \neq \phi$.

Near 2. $x \in A \Rightarrow x \nu A$.

Near 3. $x \nu (A \cup B) \Rightarrow x \nu A$ or $x \nu B$.

Near 4. $x \nu A$ and $A \subset B \Rightarrow x \nu B$.

We will begin our study of topology by stripping away all of the structure of euclidean space except ν and the four properties which we will take as axioms.

Definition A *nearness space* is a pair (X, ν) where X is a set and ν is a *nearness relation* on X, i.e., a relation between points of X and subsets of X satisfying the properties (or axioms) Near 1 to Near 4. Often we will abuse notation by suppressing ν and talking of a nearness space X.

There are many examples of nearness spaces including, of course, $\mathbb{R}^n$ with the (standard) nearness relation defined by use of pythagorean distance above. Whenever you have difficulty understanding a particular concept, restrict your attention to the case where X is a subset of $\mathbb{R}^n$ (or $\mathbb{R}^3$ or $\mathbb{R}^2$ or $\mathbb{R}$) and ν is the nearness relation above.

Examples Let X be any set and define nearness relations ν_d and ν_c on X as follows. If $x \in X$ and $A \subset X$, define $x \nu_d A$ iff $x \in A$ and $x \nu_c A$ iff $A \neq \phi$. The pair (X, ν_d) is called the *discrete* space while the pair (X, ν_c) is called the *concrete* or *indiscrete* space.

Using Fig. 2 as our guide, we can try to draw pictures representing these two spaces. In (X, ν_d), points never cluster close together as they do in $\mathbb{R}^2$. In particular, if $x \in A$ then $x \not\nu_d (A - \{x\})$, so that x bears the same relationship to $A - \{x\}$ as y does to A in Fig. 2. Thus Fig. 3 gives the only reasonable kind of picture of (X, ν_d). This also explains the source of the name "discrete". Contrast this situation with the situation in the concrete space. As long as $A \neq \phi$, $x \nu_c A$, so that we would never

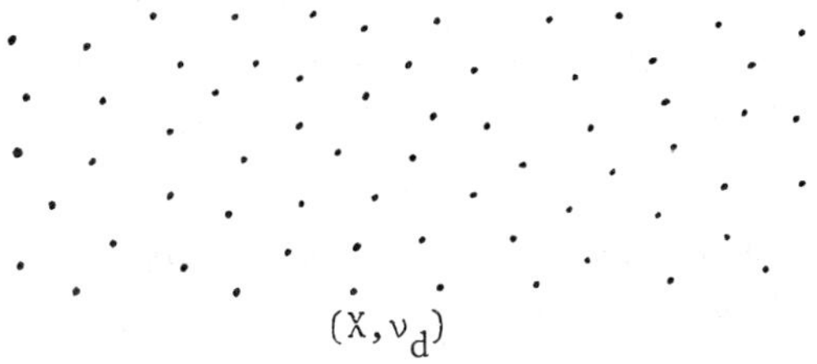

1. What Is Topology?

(X, ν_d)

FIGURE 3

have the kind of situation illustrated by y and A of Fig. 2. In particular, if we draw a picture of A then chop A in two (nonempty) pieces, each point of one piece is near the other piece. The result is that all points of X are packed tightly together in a dense mass as in Fig. 4. Hence the name "concrete".

Discrete and concrete spaces are the most extreme examples. There are many others between (in addition to $\mathbb{R}^n$). The *cofinite* nearness space is one; let X be any set and define ν on X by $x \nu A$ iff A is infinite or $x \in A$. We can define the *cocountable* nearness space by replacing "infinite" by "uncountable". We will not have much use for these two spaces.

The nearness relation defined on $\mathbb{R}^n$ above will be called the *usual* nearness relation. Unless otherwise stated, we will use the usual nearness relation on $\mathbb{R}$, $\mathbb{R}^n$, and their subsets.

(X, ν_c)

FIGURE 4

Let (X,ν) be a nearness space. By a *subspace* we mean a pair (Y,μ) where $Y \subset X$ and μ is the restriction of ν to points and subsets of Y. (Y,μ) is also a nearness space. Usually we will talk of the subspace (Y,ν) or just Y since no confusion should arise.

It is a common procedure in mathematics that when we impose a certain structure on sets we study only those functions which somehow preserve the structure: the homomorphisms of group and ring theory, the linear transformations of linear algebra, the differentiable functions of calculus, etc. We now do the same thing here.

Definition Let (X,ν) and (Y,μ) be any two nearness spaces and $f : X \to Y$ a function. Say that f *is continuous at* $x \in X$ iff for all $A \subset X$, $x \, \nu \, A \Rightarrow f(x) \, \mu \, f(A)$. Say that f is *continuous* iff f is continuous at x for all $x \in X$.

As in other situations, we often use the same symbol for the two nearness relations as long as there is no danger of confusion.

Examples When $\mathbb{R}$ has the usual nearness relation, the familiar continuous functions of elementary calculus (polynomials, sin, cos, exp, etc.) are continuous. In fact, our definition of continuous is equivalent in this context to the elementary calculus definition (see Fig. 5).

If X is a discrete space then regardless of the space Y any function $f : X \to Y$ is continuous. Similarly, if Y is a concrete space then any function $f : X \to Y$ is continuous. If Y is a subspace of X then the inclusion function $Y \hookrightarrow X$ which sends $y \in Y$ to $y \in X$ is continuous. The restriction of a continuous function to a subspace is also continuous.

THEOREM 1. Let X, Y, and Z be nearness spaces and $f : X \to Y$ and $g : Y \to Z$ continuous functions. Then $gf : X \to Z$ is also continuous.

Proof: Trivial. □

1. What Is Topology?

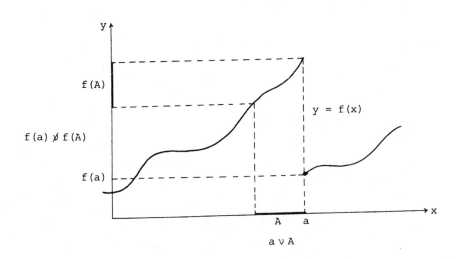

FIGURE 5

Using properties of $\mathbb{R}$ and $\mathbb{R}^n$, one can prove other familiar standard facts about combinations of continuous functions with range $\mathbb{R}$ or $\mathbb{R}^n$, e.g., if $f,g : X \longrightarrow \mathbb{R}^n$ are continuous (some nearness relation on X, usual nearness relation on $\mathbb{R}^n$) then so are $f \pm g$; if $f,g : X \longrightarrow \mathbb{R}$ are continuous then so is $f \times g$, etc.

A special kind of continuous function, the topological analogue of isomorphism, is singled out. Recall that a function $f : X \longrightarrow Y$ is an injection if for all $x,y \in X$, $f(x) = f(y) \Rightarrow x = y$, is a surjection if for all $y \in Y$, there exists $x \in X$ with $f(x) = y$, and is a bijection if it is both an injection and a surjection. If $f : X \longrightarrow Y$ is a bijection then f has a unique inverse function, denoted $f^{-1} : Y \longrightarrow X$.

Definition Suppose X and Y are nearness spaces. A function $f : X \longrightarrow Y$ is called a *homeomorphism* iff f is a bijection and both $f : X \longrightarrow Y$ and $f^{-1} : Y \longrightarrow X$ are continuous. If there is a homeomorphism between two nearness spaces, we say that they are *homeomorphic*.

Homeomorphic spaces are topologically indistinguishable. Statement V on page 1 can be made more precise by saying that the coffee cup and the doughnut are homeomorphic.

Examples Define $h : (-1,1) \to \mathbb{R}$ by $h(x) = x/(1 - |x|)$. Giving $(-1,1)$ and $\mathbb{R}$ the usual nearness relation makes h into a homeomorphism. The function $t : (-1,1) \to \mathbb{R}$ defined by $t(x) = \tan(\pi x/2)$ is also a homeomorphism (usual nearness). On the other hand, if we consider $S^1 = \{x \in \mathbb{R}^2 \mid |x| = 1\}$ to have the nearness relation inherited as a subspace of $\mathbb{R}^2$, then the function $f : [0,2\pi) \to S^1$ defined by $f(x) = (\cos x, \sin x)$ is a continuous bijection but not a homeomorphism since f^{-1} is not continuous at $(1,0)$. In fact there is no homeomorphism between the spaces $[0,2\pi)$ and S^1. Let C and Z be the two following subspaces of $\mathbb{R}^2$ with the usual nearness relation.

$$C = \{(x,y) \in \mathbb{R}^2 \mid x^2 + y^2 = 1 \text{ and } x \leq \tfrac{1}{2}\}$$
$$Z = \{(x,y) \in \mathbb{R}^2 \mid |x| \leq 1 \text{ and either } |y| = 1 \text{ or } y = x\}$$

The two nearness spaces C and Z are homeomorphic (draw pictures!).

A property of nearness spaces is called a *topological property* or *topological invariant* iff whenever it is possessed by one nearness space it is also possessed by all other homeomorphic nearness spaces. This gives meaning to statement VI on page 1, although to really understand the meaning we must do some topology.

Examples "Finite," "infinite," and "uncountable" are clearly topological invariants, although not very interesting since they do not use the nearness relations. "Discrete" and "concrete" are also topological invariants. A subset X of $\mathbb{R}^n$ is *bounded* if there is a real number M such that for all $x \in X$, $|x| \leq M$. The interval $(-1,1)$ is bounded but $\mathbb{R}$ is not bounded. Since $(-1,1)$ and $\mathbb{R}$ are homeomorphic, "bounded" is not a topological invariant.

1. What Is Topology?

Connectedness is an important nontrivial example of a topological invariant. To define this notion, we use the simplest disconnected space, namely, 2, which consists of the set {0,1} with the discrete nearness relation. 2 is the prototype disconnected space. An arbitrary space is disconnected provided it can be split continuously into two separate pieces; otherwise it is connected. Continuously splitting a space into two pieces involves finding a continuous function from the space onto 2 and this is our definition.

Definition Say that the space X is *disconnected* iff there is a continuous surjection $\delta : X \longrightarrow 2$ (see Fig. 6). Call δ a *disconnection* of X. Say that X is *connected* iff every continuous function $f : X \longrightarrow 2$ is constant. A subset C of X is connected or disconnected according as the subspace determined by C is.

THEOREM 2. Let $f : X \longrightarrow Y$ be continuous and let C be a connected subset of X. Then f(C) is a connected subset of Y.

Proof: If f(C) is not connected, then there is a disconnection $\delta : f(C) \longrightarrow 2$. Since $f : X \longrightarrow Y$ and $\delta : f(C) \longrightarrow 2$ are continuous, the composition $\delta f|C : C \longrightarrow 2$ must be continuous. Clearly δf is also a surjection. Thus C is disconnected, a contradiction. □

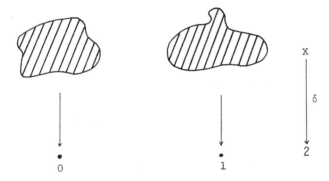

FIGURE 6

COROLLARY 3. Connectedness is a topological invariant.

Proof: Suppose X is connected and h : X ⟶ Y is a homeomorphism. Then Y = h(X) is connected, by Theorem 2. □

THEOREM 4. Let X be a nearness space and let $x \in X$. Let C be a collection of connected subsets of X each of which contains x. Then the union of all members of C is connected.

Proof: Let D denote the union of all members of C and suppose f : D ⟶ 2 is continuous. We claim that for all $y \in D$, $f(y) = f(x)$. Indeed, if $y \in D$, then there exists $C \in C$ with $y \in C$. Since C is connected, $f|C$ is constant. Thus, since $x, y \in C$, we have $f(x) = f(y)$, as claimed.

The claim of the previous paragraph implies that f is constant, and hence that D is connected. □

According to Theorem 4, each point of a nearness space is contained in a unique maximal connected set, viz., the union of all connected sets containing the point. "Maximal" here means that any connected set containing the maximal set is equal to it. A maximal connected set is called a *component*. Two components are either the same or, by Theorem 4, disjoint.

We complete this chapter by characterizing the connected subsets of $\mathbb{R}$. They are precisely the intervals.

Definition A subset I of $\mathbb{R}$ is an *interval* iff for all $a, b \in I$ and all $c \in \mathbb{R}$ with $a \leq c \leq b$, we have $c \in I$.

Intervals are of the form (a,b), or $(a,b]$, or $[a,b)$, or $[a,b]$, with $-\infty \leq a \leq b \leq \infty$. [Of course, not all of these intervals exist for all choices of a and b, e.g., $[-\infty, \infty)$.]

The next theorem requires the completeness axiom for $\mathbb{R}$. Say that the number b is an *upper bound* for the subset $X \subset \mathbb{R}$ if for all $x \in X$, $x \leq b$. If there is an upper bound for X then we say that X is *bounded above*. A *least upper bound*, say β, for X is an upper bound for X for which $\beta \leq b$ whenever b is any other upper

1. What Is Topology?

bound for X. No set has more than one least upper bound. The completeness axiom for $\mathbb{R}$ asserts that every nonempty subset of $\mathbb{R}$ which is bounded above has a least upper bound.

THEOREM 5. Let $\mathbb{R}$ have the usual nearness relation. Then a subset A of $\mathbb{R}$ is connected iff A is an interval.

Proof: Suppose A is not an interval. Then there exist $a,b,c \in \mathbb{R}$ with $a < c < b$, $a,b \in A$ but $c \notin A$. Define $\delta : A \to 2$ by $\delta(x) = 0$ if $x < c$ and $\delta(x) = 1$ if $x > c$. Then δ is continuous (why?) and is a surjection, hence A is not connected. Thus if A were connected then A would be an interval.

Conversely, suppose A is an interval but is not connected. We obtain a contradiction. Let $\delta : A \to 2$ be a disconnection of A, say $a,b \in A$ satisfy $\delta(a) = 0$, $\delta(b) = 1$. We may assume that $a < b$. Consider

$$B = \{x \in A \mid x < b \text{ and } \delta(x) = 0\}$$

The reader should draw a picture of B to guide him through the rest of the proof. Now $a \in B$ and b is an upper bound for B. Hence by the completeness axiom for $\mathbb{R}$, B has a least upper bound, say β. Now A is an interval and $a \leq \beta \leq b$, so $\beta \in A$. Our contradiction is obtained by seeing where δ takes β.

On one hand, since β is the *least* upper bound of B, $\beta \vee B$, so $\delta(\beta) \vee \delta(B)$, i.e., $\delta(\beta) \vee \{0\}$. Since 2 has the discrete nearness relation we must have $\delta(\beta) = 0$. On the other hand, since for all $x \in (\beta, b]$, $\delta(x) = 1$, and (unless $\beta = b$) $\beta \vee (\beta, b] \subset A$, we must have $\delta(\beta) \vee \delta((\beta, b])$, i.e., $\delta(\beta) \vee \{1\}$, so $\delta(\beta) = 1$ (even if $\beta = b$!). This gives us the desired contradiction.

Thus if A is an interval then A is connected. □

Note that there is no simple analogue of Theorem 5 in higher dimensions. For example, the subsets $\{(x,y) \in \mathbb{R}^2 \mid x = 0$ and $-1 \leq y \leq 1\}$ and $\{(x,y) \in \mathbb{R}^2 \mid x > 0$ and $y = \sin(1/x)\}$ of $\mathbb{R}^2$, being homeomorphic to $[-1,1]$ and $(0,\infty)$, respectively, are connected. Their union, depicted in Fig. 7, is also connected.

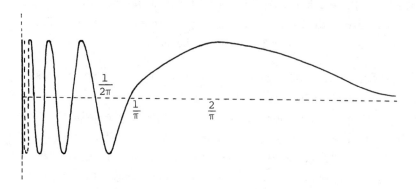

FIGURE 7

EXERCISES

1. For the following pairs (x,A), where $x \in \mathbb{R}^2$, $A \subset \mathbb{R}^2$, determine whether $x \vee A$. Explain your answers.
   (a) $x = 0$, $A = \{(x_1,x_2) \mid x_1 x_2 > 0\}$
   (b) $x = 0$, $A = \{(x_1,x_2) \mid x_1 \in \mathbb{Z}, x_2 \in \mathbb{Z}, x_1^2 + x_2^2 \neq 0\}$
   (c) $x = 0$, $A = \{(x_1,x_2) \mid x_1 \in \mathbb{Q}, x_2 \in \mathbb{Q}, x_1^2 + x_2^2 \neq 0\}$
   (d) $x = 0$, $A = \{(x_1,x_2) \mid \text{at least one } x_i \text{ is irrational}\}$
   (e) $x = (1,0)$, $A = \{(x_1,x_2) \mid x_1^2 + x_2^2 < 1\}$
   (f) $x = (1,0)$, $A = \{(x_1,x_2) \in S^1 \mid x_2 < 0\}$
   (g) $x = 0$, $A = \{(x_1,x_2) \mid x_1 > 0, x_2 = \sin(1/x_1)\}$

2. Verify that the usual nearness relation on $\mathbb{R}^n$, the discrete, concrete, and cofinite nearness relations all satisfy the axioms Near 1 to Near 4.

3. Write down all possible nearness relations on the set $\{0,1\}$. Identify which pairs give rise to homeomorphic nearness spaces.

4. Verify that the function $g : S^1 \longrightarrow [0,2\pi)$, which is the inverse of the function $f : [0,2\pi) \longrightarrow S^1$ defined by $f(x) = (\cos x, \sin x)$, is not continuous at $(1,0)$.

5. Let $f : X \longrightarrow Y$ be a function between two nearness spaces. Suppose $X = X_1 \cup X_2$, where the two subsets X_1 and X_2 satisfy $x \vee X_i \Rightarrow x \in X_i$. Prove that if the restrictions $f \mid X_i : X_i \longrightarrow Y$ are continuous then so is f.

1. What Is Topology?

6. Suppose $f,g : X \to \mathbb{R}$ are continuous at $x \in X$, and $A \subset X$ is such that $x \, \nu \, A$. Thus $f(x) \, \nu \, f(A)$ and $g(x) \, \nu \, g(A)$. Prove that for all $r > 0$, there exists $a \in A$ with $|f(x) - f(a)| < r$ and $|g(x) - g(a)| < r$. [Thus the same a does for both $f(x) \, \nu \, f(A)$ and $g(x) \, \nu \, g(A)$.]

7. (a) Define the functions $s,p : \mathbb{R}^2 \to \mathbb{R}$ by $s(x,y) = x + y$ and $p(x,y) = xy$. Verify that s and p are continuous.
   (b) Prove that if $f,g : X \to \mathbb{R}$ are continuous, where X is any nearness space, then so are $f + g$ and $f \times g$.

8. Prove that discreteness and concreteness are topological invariants.

9. Prove that the function $h : (-1,1) \to \mathbb{R}$ given by $h(x) = x/(1-|x|)$ is a homeomorphism.

10. Determine which of the following subsets of $\mathbb{R}$, $\mathbb{R}^2$, or $\mathbb{R}^n$ are connected. Justify your answers.
    (a) Subsets of $\mathbb{R}$: ϕ; $\mathbb{R}$; $\{0\} \cup \{1/k \mid k = 1, 2, \ldots\}$; $\{x \in \mathbb{Q} \mid 0 \leq x \leq 1\}$.
    (b) Subsets of $\mathbb{R}^2$: S^1; $\{(x_1,x_2) \mid x_1 x_2 \geq 0\}$; $\{(x_1,x_2) \mid x_1 = 0 \text{ or } x_2 = 0 \text{ or } x_1 x_2 = 1\}$; $\{(x_1,x_2) \mid x_1 = 0 \text{ and } -1 \leq x_2 \leq 1\} \cup \{(x_1,x_2) \mid x_1 > 0 \text{ and } x_2 = \sin(1/x_1)\}$.
    (c) Subsets of $\mathbb{R}^n$: S^n; the x_1 axis in $\mathbb{R}^n$.

11. Determine all connected subsets of: (i) the discrete space; (ii) the concrete space. (Hint: Refer to Figs. 3 and 4).

2
TOPOLOGICAL SPACES

The aim of this chapter is to put topology in its usual setting. This consists of a certain collection of subsets of a space, the open subsets. The pair consisting of a set and a collection of open subsets (satisfying certain conditions) is called a topological space. Although slightly less general, the topological space is a much neater and more economical structure than the nearness space and we will use it most of the time. In fact open sets form the foundation on which we will build our differential structures and many times in later chapters we will refer back to open sets as the basis of a construction. Another type of set introduced in this chapter is the neighborhood. A neighborhood of a point is, loosely speaking, a set containing all points in the vicinity. This kind of set, too, is important. For example, differentiability of a function at a point is entirely determined in a neighborhood of the point. This we know from elementary calculus and we will find to be the case in a much more general setting. Despite the above, nearness spaces will be convenient to use from time to time, particularly in view of their geometric appeal.

Definition Let (X,ν) be a nearness space and let $x \in X$, $N \subset X$. Say that N is a *neighborhood* of x iff $x \not\nu X-N$. Essentially, a neighborhood of x contains all points of X in the vicinity of x. In particular, x lies in every neighborhood of x.

A subset U of X is *open* iff for all $x \in U$, $x \not\nu X-U$. Thus U is open iff it is a neighborhood of each of its points.

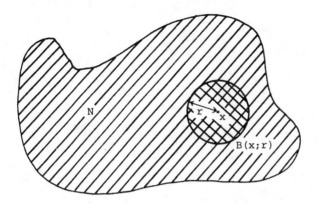

FIGURE 8

Examples In $\mathbb{R}^n$ with the usual nearness relation, N is a neighborhood of x iff there exists $r > 0$ such that $B(x;r) \subset N$ (see Fig. 8), where

$$B(x;r) = \{y \in \mathbb{R}^n \mid |x-y| < r\}$$

is the ball centered at x of radius r. In particular, $\mathbb{R}^n$ is itself a neighborhood of each of its points, so is open. Let $B^n = \{x \in \mathbb{R}^n \mid |x| \leq 1\}$. Then $B^n = B(0;1) \cup S^{n-1}$, where

$$S^{n-1} = \{x \in \mathbb{R}^n \mid |x| = 1\}$$

The "closed" ball B^n is a neighborhood of each point of $B(0;1)$ but not of any point of S^{n-1}. The open ball $B(0;1)$ is open but neither B^n nor S^{n-1} is.

THEOREM 1. Let (X,ν) be a nearness space. Then the open subsets of X satisfy the following:

Open 1. ϕ is open.

Open 2. X is open.

Open 3. If U and V are open then so is $U \cap V$.

Open 4. If $\{U_\alpha \mid \alpha \in A\}$ is a (possibly infinite) family of open sets then $\bigcup_{\alpha \in A} U_\alpha$ is open.

2. Topological Spaces

Proof: Open 1. An immediate consequence of the definition.

Open 2. Taking $U = X$ in the definition, we must show that for all $x \in X$, $x \not\nu\ X - X$. But $X - X = \phi$, so the criterion is satisfied by Near 1.

Open 3. Given $x \in U \cap V$, we must show that $x \not\nu\ X - (U \cap V)$. Suppose instead that $x \nu\ X - (U \cap V)$. Now $X - (U \cap V) = (X - U) \cup (X - V)$, so by Near 3, we must have either $x \nu\ X - U$ or $x \nu\ X - V$. But then either U or V is not open.

Open 4. Given $x \in \bigcup_{\alpha \in A} U_\alpha$, say $x \in U_\beta$, suppose that $x \nu\ X - \bigcup_{\alpha \in A} U_\alpha$. Now $X - \bigcup_{\alpha \in A} U_\alpha \subset X - U_\beta$, so by applying Near 4 to the point x and the sets $X - \bigcup_{\alpha \in A} U_\alpha$ and $X - U_\beta$, we deduce that $x \nu\ X - U_\beta$ even though $x \in U_\beta$. This contradicts the openness of U_β. □

Remark Using induction and Open 3, we can deduce that a finite intersection of open sets is open. However, an arbitrary intersection of open sets need not be open. For example, let $U_n = \{x \in \mathbb{R}\ |\ |x| < 1/n\}$, $n = 1, 2, \ldots$. Then for each n, U_n is open in $\mathbb{R}$ but $\bigcap_{n=1}^{\infty} U_n = \{0\}$ is not open.

Definition A *topological space* is a pair (X, T), where X is a set and T is a collection of subsets of X, called *open* sets, satisfying Open 1 to Open 4 above. The collection T is called a *topology* (on X).

Examples Let X be a set. Then the discrete and concrete nearness relations on X give rise to the *discrete* and *concrete topologies* respectively. The discrete topology consists of all subsets of X and the concrete topology consists of only ϕ and X. The *usual topology* on $\mathbb{R}^n$ consists of all sets U satisfying the condition

$\forall\ x \in U$, $\exists\ r > 0$ such that $B(x;r) \subset U$

Unless otherwise stated, we will use the usual topology on $\mathbb{R}$, $\mathbb{R}^n$, and their subsets. The cofinite nearness space gives rise to the *cofinite topology* in which a set is open iff it is ϕ or has a finite complement. Hence the name "cofinite".

The two concepts, nearness space and topological space, are closely related in the following sense. Let X be a set, let N be the collection of all nearness relations on X, and T be the collection of all topologies on X. Then there are two functions $\alpha : N \longrightarrow T$ and $\beta : T \longrightarrow N$ with $\alpha\beta$ the identity.

The function α has already been defined above, i.e., given a nearness relation ν on X, $\alpha(\nu)$ is the collection of all open sets defined by means of ν. Conversely, given a topology T on X, define the nearness relation $\beta(T)$ on X by x $\beta(T)$ A iff for all U $\in T$ with x $\in$ U, U $\cap$ A $\neq \phi$. It can be checked that $\beta(T)$ is a nearness relation, and that $\alpha\beta$ is the appropriate identity.

In view of the above, any procedure which can be carried out on a nearness space can be transferred to a topological space, and any procedure which can be carried out on a topological space can be transferred to those nearness spaces of the form $(X,\beta(T))$. We have already seen how discrete and concrete nearness relations give rise to discrete and concrete topologies. Similarly the definitions of continuous function and connected space carry over to topological spaces. The former is done in Theorem 4 below. Corollary 5 gives a characterization of connectedness involving only the topology, with no reference to continuous functions.

Definition Let (X,T) be a topological space. A subfamily B of T is called a *base* or *basis* for T provided every member of T is a union of members of B. Usually one specifies a topology by describing a basis, so the following criterion enables one to check whether an alleged basis really is a basis for a topology.

PROPOSITION 2. Let F be a family of subsets of a set X. Then F is a basis for a topology on X iff UF = X and for all A,B $\in F$ and all x $\in$ A $\cap$ B, there exists C $\in F$ such that x $\in$ C $\subset$ A $\cap$ B.

Proof: Suppose F is a basis for the topology T on X. Thus

T = {UG | $G \subset F$}

Since X $\in T$, there exists $G \subset F$ with X = UG. Of course

2. Topological Spaces

$UG \subset UF \subset X$, so $UF = X$. Let $A, B \in F$ and $x \in A \cap B$. Since $F \subset T$ and T is a topology, we have $A \cap B \in T$. Thus there exists $G \subset F$ with $A \cap B = UG$. In particular, there exists $C \in G$ with $x \in C \subset A \cap B$. Of course $C \in F$.

Conversely, suppose F satisfies the criterion. Let

$$T = \{UG \mid G \subset F\}$$

We need to verify that T is a topology on X. Since $\phi \subset F$ and $F \subset F$, by definition, $U\phi \in T$ and $UF \in T$, i.e., $\phi \in T$ and $X \in T$, which verifies Open 1 and Open 2. Let $G_1, G_2 \subset F$, so that UG_1 and UG_2 are two typical members of T. To verify Open 3, we must find $H \subset F$ so that $UH = (UG_1) \cap (UG_2)$. Let $x \in (UG_1) \cap (UG_2)$. Then there exists $G_i \in G_i$ such that $x \in G_i$, $i = 1, 2$. Since $G_1, G_2 \in F$ and $x \in G_1 \cap G_2$, by the condition on F, there exists $H_x \in F$ such that $x \in H_x \subset G_1 \cap G_2 \subset (UG_1) \cap (UG_2)$. Thus for all $x \in (UG_1) \cap (UG_2)$, we can find $H_x \in F$ such that $x \in H_x \subset (UG_1) \cap (UG_2)$. Let $H = \{H_x \mid x \in (UG_1) \cap (UG_2)\}$. Then $H \subset F$, so $UH \in T$. Moreover, each member of H lies in $(UG_1) \cap (UG_2)$, so $UH \subset (UG_1) \cap (UG_2)$. On the other hand, each point of the latter set lies in some member of H, so the two sets are equal. This verifies Open 3. Open 4 is obvious. □

Examples The collection of open intervals forms a basis for the usual topology on $\mathbb{R}$. More generally, $\{B(x;r) \mid x \in \mathbb{R}^n, r > 0\}$ forms a basis for the usual topology on $\mathbb{R}^n$. For any set X, the family of all singleton sets forms a basis for the discrete topology.

Neighborhoods provide a third approach to topological spaces. Although we will not fully develop this approach, we will give the basic properties of neighborhoods. Note that they can be defined directly from open sets: N is a neighborhood of x iff there is an open set U with $x \in U \subset N$.

THEOREM 3. Let X be a nearness or topological space, and let $x \in X$. Then the family of neighborhoods of x satisfies the following:

Nbd 1. If N is a neighborhood of x then $x \in N$.

Nbd 2. If N is a neighborhood of x and $N \subset A \subset X$ then A is a neighborhood of x.

Nbd 3. If N_1 and N_2 are neighborhoods of x then so is $N_1 \cap N_2$.

Nbd 4. If N is a neighborhood of x then there is a neighborhood A of x such that $A \subset N$ and A is a neighborhood of all of its points.

Proof: Proofs can be constructed directly from the nearness axioms or from the openness axioms. We will give the latter.

Nbd 1. This follows from the definition.

Nbd 2. This follows from the definition.

Nbd 3. If N_1 and N_2 are neighborhoods of x then by the definition, there are open sets U_1 and U_2 such that $x \in U_i \subset N_i$. Now by Open 3, $U_1 \cap U_2$ is open. Thus, since $x \in U_1 \cap U_2 \subset N_1 \cap N_2$, $N_1 \cap N_2$ is also a neighborhood of x.

Nbd 4. Let $A = \cup \{U \mid U$ is open and $U \subset N\}$. Since N is a neighborhood of x, A is also a neighborhood of x. Clearly $A \subset N$. By Open 4, A is open, so it is a neighborhood of each of its points. □

Definition Let (X, T) be a topological space. A subset C of X is *closed* iff $X - C \in T$. For any subset $Y \subset X$, we define three sets: Int Y (the *interior* of Y), Cl Y (the *closure* of Y) and Fr Y (the *frontier* of Y) by the formulas

Int $Y = \cup \{U \in T \mid U \subset Y\}$

Cl $Y = \cap \{C \subset X \mid C$ is closed and $Y \subset C\}$

Fr $Y = $ Cl $Y - $ Int Y

Note that Int Y is open and Cl Y and Fr Y are both closed. The nearness relation constructed from the topology is defined by $x \vee A$ iff $x \in $ Cl A. A is closed iff $x \vee A \Rightarrow x \in A$. This criterion will be used in the proof of Theorem 4 below.

Warning "Closed" does *not* mean "not open." In fact, there are topologies in which some sets are both open and closed and some sets are neither open nor closed. Since X is always open,

2. Topological Spaces

$\phi = X - X$ is always closed; since ϕ is always open, $X = X - \phi$ is always closed. Thus ϕ and X are always both open and closed. In the concrete space all other subsets are neither open nor closed. In the discrete space all subsets are both open and closed. In $\mathbb{R}^n$ with the usual topology, some sets, e.g., $B^n - \{0\}$, are neither open nor closed, most open sets are not closed, most closed sets are not open, and ϕ and $\mathbb{R}^n$ are both open and closed (and they are the only such sets; cf. Corollary 5 below).

THEOREM 4. Let $f : X \longrightarrow Y$ be a function between two topological spaces. Then conditions a and b below are equivalent and conditions c, d, e, and f are equivalent.

(a) f is continuous at x.
(b) For every neighborhood V of f(x) in Y, there is a neighborhood U of x in X such that $f(U) \subset V$.
(c) f is continuous.
(d) For every closed subset C of Y, $f^{-1}(C)$ is closed in X.
(e) If B is a basis for the topology of Y, then for every $V \in B$, $f^{-1}(V)$ is open in X.
(f) For every open subset V of Y, $f^{-1}(V)$ is open in X.

  Proof: (a) $\Rightarrow$ (b): If f is continuous at x and V is a neighborhood of f(x) in Y, let $U = f^{-1}(V)$. Then $f(U) = ff^{-1}(V) \subset V$, so it remains to show that U is a neighborhood of x. Suppose not. Then $x \, \nu \, X-U$, so by continuity, $f(x) \, \nu \, f(X-U)$. But $f(X-U) \subset Y - V$. Thus by Near 4, $f(x) \, \nu \, Y-V$ and hence V is not a neighborhood of $f(x)$.

  (b) $\Rightarrow$ (a): Suppose that $A \subset X$ and $x \, \nu \, A$. We must show that $f(x) \, \nu \, f(A)$. Suppose instead that $f(x) \not\nu f(A)$. Then $Y - f(A)$ is a neighborhood of $f(x)$, so there is a neighborhood U of x in X such that $f(U) \subset Y - f(A)$. Thus $x \not\nu X-U$ and $U \cap A = \phi$, i.e., $A \subset X - U$. By Near 4, we have $x \not\nu A$, a contradiction.

  (c) $\Rightarrow$ (d): Let C be closed in Y and suppose $x \, \nu \, f^{-1}(C)$. Using the closed sets criterion above, it suffices to show that $x \in f^{-1}(C)$. By c, $f(x) \, \nu \, ff^{-1}(C)$, so by Near 4 $f(x) \, \nu \, C$ since $ff^{-1}(C) \subset C$. Thus, since C is closed, $f(x) \in C$, which is the same as $x \in f^{-1}(C)$.

(d) ⇒ (e): Suppose B is a basis for the topology of Y and $V \in B$. Then V is open in Y, so $Y - V$ is closed; hence by d, $X - f^{-1}(V) = f^{-1}(Y - V)$ is closed, so that $f^{-1}(V)$ is open.

(e) ⇒ (f): Suppose B is a basis for Y and V is an open subset of Y. Then for some subfamily F of B, we have $V = \cup F$. Now for all $F \in F$, the set $F \in B$, so by e, $f^{-1}(F)$ is open in X. But $f^{-1}(V) = \cup \{f^{-1}(F) \mid F \in F\}$, so by Open 4, $f^{-1}(V)$ is open.

(f) ⇒ (c): The equivalence of a and b is used. Let $x \in X$, and suppose V is a neighborhood of $f(x)$ in Y. By Nbd 4, there is an open neighborhood A of $f(x)$ so that $A \subset V$. By f, $U = f^{-1}(A)$ is open. Further, $x \in U$, so U is a neighborhood of x. Note that $f(U) = ff^{-1}(A) \subset A \subset V$. □

Remarks 1. Condition b of Theorem 4 reduces to the familiar ε, δ definition of elementary calculus when we use the usual topology on $\mathbb{R}$. This follows from the fact that V is a neighborhood of x in $\mathbb{R}$ iff there exists $\varepsilon > 0$ such that $(x - \varepsilon, x + \varepsilon) \subset V$.

2. There was a slight ambiguity in the interpretation of criterion e in the proof above. In order to verify continuity using e, we need only verify that $f^{-1}(V)$ is open for all V in some particular basis for the topology on Y. A consequence of the proof is that if this is the case then $f^{-1}(V)$ is open for all V in any basis for the topology on Y.

COROLLARY 5. A space X is connected iff the only subsets of X which are both open and closed are ϕ and X.

Proof: Suppose $A \subset X$ is both open and closed yet $\phi \neq A \neq X$. Define $\delta : X \rightarrow 2$ by $\delta(A) = \{0\}$ and $\delta(X - A) = \{1\}$. Since $\{\{0\},\{1\}\}$ forms a basis for the (discrete) topology on 2, and $\delta^{-1}(\{0\}) = A$ and $\delta^{-1}(\{1\}) = X - A$ are both open in X, by Theorem 4, δ is continuous. Thus, as δ is surjective, δ is a disconnection of X. Contrapositively, if X is connected, then its only open and closed subsets are ϕ and X.

Conversely, if $\delta : X \rightarrow 2$ is a disconnection of X, then $\delta^{-1}(\{0\})$, being the inverse image of an open and closed set, is open

2. Topological Spaces

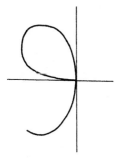

FIGURE 9

and closed. Further, $\delta^{-1}(\{0\})$ is neither ϕ nor X because δ is surjective. Thus for ϕ and X to be the only open and closed subsets of X, the space X must be connected. □

Definition A function $f : X \longrightarrow Y$ is an *embedding* provided it is a homeomorphism from X to $f(X)$. An embedding satisfies all of the requirements for a homeomorphism except surjectivity. For many purposes it is convenient to identify X and $f(X)$ when $f : X \longrightarrow Y$ is an embedding, i.e., to treat X as a subspace of Y.

Examples Inclusion functions are embeddings. $e : (-1,1) \longrightarrow \mathbb{C}$ defined by $e(t) = e^{\pi i t}$ is an embedding (usual topologies). The function $f : (5/4, 2) \longrightarrow \mathbb{C}$ defined by

$$f(t) = e^{\pi i t} \sin 2\pi t$$

is not an embedding even though it is continuous and injective. As suggested by Fig. 9, the inverse function is not continuous at the complex number $f(3/2) = 0$.

EXERCISES

1. Determine whether the following subsets of $\mathbb{R}$ or $\mathbb{R}^n$ (usual topologies) are open or closed. Find the interior, closure, and frontier of each set.

$$A = \phi$$
$$B = \mathbb{R}$$
$$C = [0,1] \subset \mathbb{R}$$
$$D = (0,1) \subset \mathbb{R}$$
$$E = [-184,405] \cup \{1000\} \subset \mathbb{R}$$
$$F = (-184,405) \cup \{1000\} \subset \mathbb{R}$$
$$G = (-\infty,1000) \cup (1000,\infty) \subset \mathbb{R}$$
$$H = S^{n-1} = \{x \in \mathbb{R}^n \mid |x| = 1\}$$
$$I = \bigcup_{i=1}^{\infty} \left\{ x \mid \frac{1}{2i} < |x| < \frac{1}{2i-1} \right\} \subset \mathbb{R}^n$$
$$J = \{0\} \cup \bigcup_{i=1}^{\infty} \left\{ x \mid \frac{1}{2i} \leq |x| \leq \frac{1}{2i-1} \right\} \subset \mathbb{R}^n$$

2. For the following subsets N of $\mathbb{R}^n$, determine whether N is a neighborhood of 0. Give $\mathbb{R}^n$ the usual topology.
    (a) $N = \{x \mid |x| < 1\}$
    (b) $N = \{x \mid |x| \leq 1\}$
    (c) $N = \{0\} \cup \{(x_1,\ldots,x_n) \mid x_1 x_2 \cdots x_n \neq 0\}$
    (d) $N = \{0\} \cup \{(x_1,\ldots,x_n) \mid$ at least one x_i is irrational$\}$
    (e) $N = \{0\} \cup \bigcup_{i=1}^{\infty} \left\{ x \mid \frac{1}{2i} < |x| < \frac{1}{2i-1} \right\}$

3. Verify directly (i.e., without referring to the nearness axioms) that the usual topology on $\mathbb{R}^n$ and the discrete and concrete topologies satisfy the axioms Open 1 to 4.

4. Show that the cofinite topology on X is the same as the discrete topology iff X is finite.

5. Verify that if $Y \subset X$ then $\text{Cl } Y = X - \text{Int } (X - Y)$.

6. (a) Let $(X,\mathcal{T})$ and $(Y,\mathcal{U})$ be topological spaces. Verify that $\{T \times U \subset X \times Y \mid T \in \mathcal{T} \text{ and } U \in \mathcal{U}\}$ is a basis for a topology on $X \times Y$. This topology is called the *product topology* on $X \times Y$.
    (b) Verify that the usual topology on $\mathbb{R}^2$ is the product topology on $\mathbb{R} \times \mathbb{R}$ obtained from the usual topology on $\mathbb{R}$.

2. Topological Spaces

(c) Suppose $f_i : X_i \to Y_i$ is continuous for $i = 1, 2$. Prove that $f_1 \times f_2 : X_1 \times X_2 \to Y_1 \times Y_2$ is continuous when $X_1 \times X_2$ and $Y_1 \times Y_2$ are given the respective product topologies.

7. Divide the following subspaces of $\mathbb{R}^2$ or $\mathbb{R}^3$ (usual topologies) into classes, where two spaces lie in the same class iff they are homeomorphic. (It seems desirable to develop one's intuition for this kind of thing: *proving* that two spaces are not homeomorphic is not necessarily easy, although that too should be tried in a few cases.)

A,B,C,D,E,F,G,H,I,J,K,L,M,N,O,P,Q,R,S,T,U,V,W,X,Y,Z,1,2,3,4,5,6,7,8,9

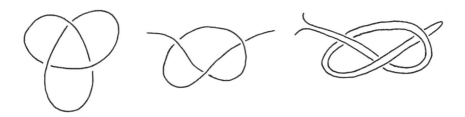

(The last three sets are meant to be contained in $\mathbb{R}^3$ and are intended to be like knotted pieces of string. In all cases the lines are meant to have zero thicknesses.)

8. Prove that for the functions $\alpha : N \to T$ and $\beta : T \to N$, where N is the collection of all nearness relations on a set X and T is the collection of all topologies on X, $\alpha\beta$ is the identity (cf. Ref. 6 in the Bibliography).

3

SOME TOPOLOGICAL PROPERTIES

There are many named topological properties, for example cactoid, uniformly locally simply connected, normal, countably metacompact, $T_{3\frac{1}{2}}$, locally peripherally compact, etc. The index of a typical topology textbook might list 40 such properties, although less than half of these would be considered important. We will consider only four important topological properties, the first of which, connectedness, we have already met in Chap. 1; the others we will study in this chapter. Later other topological properties will be introduced, but these will tend to have rather specific applicability.

Connectedness was studied in the context of nearness spaces, but, as noted in Chap. 2, the definition carries over easily to topological spaces. Our second important topological property (Hausdorffness) is a formalization of the intuitive feeling that if you look at two distinct points of $\mathbb{R}^n$ closely enough, then they are really a long way apart. (This is one reason why nearness spaces had to involve a relation between points and sets rather than points and points). In a general topological space, this is not the case; cf. our attempt in Fig. 4 at depicting the concrete space.

The third important topological property is compactness. It is harder to give an intuitive idea of this notion, but it is, in a way, a generalization of finiteness. We will use this idea many times later on when we carry out a particular construction locally, i.e., in a neighborhood of a point. Compactness will enable us to get by with performing the construction only finitely many times to convert it to a global construction, i.e., one carried out on the whole space.

The fourth important topological property is that of being a manifold. This is so important that, from Chap. 5 on, all of our spaces will be manifolds. The importance derives from the fact that locally, manifolds are like euclidean space, which is the standard arena of calculus. Now differentiability is a local phenomenon, so to talk about it, we need only a space which is locally like euclidean space, i.e., a manifold. Other local phenomena of euclidean space carry over to manifolds, and manifolds form the basis of large and very active fields of topological research.

Definition A topological space X is *Hausdorff* iff for all $x,y \in X$ with $x \neq y$, there are neighborhoods U and V of x and y with $U \cap V = \phi$ (see Fig. 10).

In general, the more open sets, the more likely the space will be Hausdorff. If (X,T) is Hausdorff and U is a topology on X with $T \subset U$, then (X,U) is also Hausdorff.

Examples Discrete topologies are always Hausdorff, as is the usual topology on $\mathbb{R}^n$. Concrete topologies are not Hausdorff (except on singleton or empty sets!), nor is the cofinite topology on an infinite set. Any subspace of a Hausdorff space is itself Hausdorff.

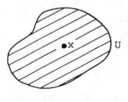
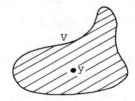

FIGURE 10

3. Some Topological Properties

THEOREM 1. Hausdorffness is a topological invariant. □

Our third topological property is compactness. This notion is much harder to understand than those of connectedness and Hausdorffness, the main reason being that the intuitive ideas behind the latter two are easy to formalize topologically, whereas the intuitive idea of compactness is not. (Exercise: Try to topologically formalize a dictionary definition of the word "compact".) Furthermore, there has been an evolution of the notion of compactness within topology, so that the present definition is rather different from the original definition. It is a very useful notion.

Definitions Let X be a set and $C \subset X$. A family F of subsets of X *covers* C or *is a cover of* C iff C is contained in the union of all members of F. A *subcover* is a subfamily of F which still covers C. If X has a topology T, then the cover F is called an *open cover* provided $F \subset T$.

Definition Say that the subset C of the topological space X is *compact* iff every open cover of C has a finite subcover. Of course, X itself is compact iff X is a compact subset of X.

In general, the fewer open sets, the more likely the space will be compact. If (X,T) is compact and U is a topology on X such that $U \subset T$, then (X,U) is also compact.

Examples Finite sets are compact. If (x_n) is a sequence in $\mathbb{R}$ which converges to x then $\{x\} \cup \{x_n \mid n = 1, 2, \ldots\}$ is compact; thus $\{0\} \cup \{1/n \mid n = 1, 2, \ldots\}$ is compact. Any bounded closed interval in $\mathbb{R}$ is compact but no nonempty open interval is. For example, to see that $(0,1)$ is not compact, consider the family

$$F = \left\{ \left(\frac{1}{n}, 1\right) \mid n = 2, 3, \ldots \right\}$$

Then F consists of open sets and covers $(0,1)$, so is an open cover of $(0,1)$. However, there is no finite subcover.

Verification of the compactness of $[0,1]$ requires the completeness axiom on $\mathbb{R}$. Suppose F is an open cover of $[0,1]$. Let

$$A = \{x \in [0,1] \mid [0,x] \text{ is covered by finitely many members of } F\}$$

Then A is nonempty (it contains 0!) and is bounded above (by 1) so by the completeness axiom it must have a least upper bound, say α. We must show that $\alpha \in A$ and $\alpha = 1$. Some member of F contains 0 and, since all members of F are open, it follows that for some $x > 0$, $[0,x]$ is contained in one member of F. Thus $\alpha > 0$. Since $\alpha \in (0,1]$, there exists $F \in F$ with $\alpha \in F$. Since F is open in $\mathbb{R}$, there exists $\varepsilon \in \mathbb{R}$ such that $0 < \varepsilon < \alpha$ and $[\alpha - \varepsilon, \alpha + \varepsilon] \subset F$. Clearly A is an interval, so $\alpha - \varepsilon \in A$ and hence $[0, \alpha - \varepsilon]$ is covered by finitely many members of F. Adding F to this collection, we see that $[0, \alpha + \varepsilon]$ is also covered by finitely many members of F. Thus $[0,\alpha]$ is covered by finitely many members of F, so $\alpha \in A$. Further, if $\alpha < 1$ then we may assume ε to be so small that $\alpha + \varepsilon \leq 1$. Thus, since $[0, \alpha + \varepsilon]$ is covered by finitely many members of F, we have $\alpha + \varepsilon \in A$, which contradicts the fact that α is an upper bound for A. Thus $\alpha = 1$. The definition of A combined with the fact that $1 \in A$ tells us that $[0,1]$ is covered by finitely many members of F.

Recall that a subset A of $\mathbb{R}^n$ is *bounded* iff there exists $M \in \mathbb{R}$ such that for all $x \in A$, $|x| \leq M$.

THEOREM 2 (Heine-Borel Theorem). A subset of $\mathbb{R}^n$ is compact iff it is closed and bounded.

Proof: Omitted, but see Exercise 7 and Appendix A. □

The Heine-Borel theorem gives us a useful characterization of compact subsets of $\mathbb{R}^n$, which will be our main arena of study. In particular, S^n is compact.

3. Some Topological Properties

THEOREM 3. Let $f : X \to Y$ be a continuous function between topological spaces and suppose C is a compact subset of X. Then $f(C)$ is a compact subset of Y.

Proof: Suppose $\mathcal{U}$ is an open cover of $f(C)$. Then by Theorem 2.4, for all $U \in \mathcal{U}$, $f^{-1}(U)$ is open. Moreover, $\{f^{-1}(U) \mid U \in \mathcal{U}\}$ is a cover of C. (Check!) Thus by compactness, there is a finite subcover, say

$$\{f^{-1}(U_i) \mid i = 1, \ldots, n\}$$

But then $\{U_i \mid i = 1, \ldots, n\}$ is a cover of $f(C)$, so that $f(C)$ is compact. □

COROLLARY 4. Compactness is a topological invariant.

Proof: Similar to Corollary 1.3. □

THEOREM 5. A closed subset of a compact space is compact; a compact subset of a Hausdorff space is closed.

Proof: If C is a closed subset of the compact space X and $\mathcal{U}$ is an open cover of C then $\mathcal{U} \cup \{X - C\}$ is an open cover of X. Let F be a finite subcover. Then $F - \{X - C\}$ is a finite subfamily of $\mathcal{U}$ covering C, so C is compact.

If C is a compact subset of the Hausdorff space X and $x \in X - C$, then for all $y \in C$, there are open sets U_y, V_y such that $x \in U_y$, $y \in V_y$, and $U_y \cap V_y = \phi$. The family $\{V_y \mid y \in C\}$ is an open cover of C; let $\{V_{y_1}, \ldots, V_{y_n}\}$ be a finite subcover. Then $U = \cap_{i=1}^{n} U_{y_i}$ is an open set containing x such that $U \cap C = \phi$. Thus $X - C$ is a neighborhood of each of its points, so is open, and hence C is closed. □

Since increasing the number of open sets increases the likelihood of Hausdorffness and decreases the likelihood of compactness, we should not be surprised to find some strong theorems involving an interplay between these two notions. Only one such theorem will be given here.

THEOREM 6. Let $f : X \to Y$ be a continuous bijection from a compact space to a Hausdorff space. Then f is a homeomorphism. Thus X is Hausdorff and Y is compact.

Proof: We must show that the function $f^{-1} : Y \to X$ is continuous. By Theorem 2.4, it is enough to show that if C is closed in X, then $(f^{-1})^{-1}(C) = f(C)$ is closed in Y. Now X is compact, so by Theorem 5, C is compact. Thus by Theorem 3, $f(C)$ is compact, hence, by Theorem 5, is closed. □

Our fourth important topological property is that of being a manifold.

Definition A *topological (m-)manifold* is a Hausdorff space M^m each point of which has an open neighborhood homeomorphic to $\mathbb{R}^m$. Our prime concern will be with compact manifolds, although we will admit noncompact ones. In the notation M^m, the letter m denotes the dimension of the manifold and is usually omitted in subsequent references to the manifold.

If $U \subset M$ is open and $\varphi : U \to \mathbb{R}^m$ is an embedding, we call (U,φ) a *(coordinate) chart*. A collection $\{(U_\alpha, \varphi_\alpha) \mid \alpha \in A\}$ of charts is called an *atlas* provided $M = \bigcup_\alpha U_\alpha$. If (U,φ) and (V,ψ) are two charts then the function

$$\psi\varphi^{-1} : \varphi(U \cap V) \to \mathbb{R}^m$$

is called a *coordinate transformation*.

It is often more convenient to deal with nonsurjective coordinate functions. Hence we require only that φ be an embedding rather than a homeomorphism. This leads to a technical problem. Later on we will want the domain of the coordinate transformation to be open. Since $U \cap V$ is open in U, then $\varphi(U \cap V)$ is open in $\varphi(U)$. To conclude that $\varphi(U \cap V)$ is open in $\mathbb{R}^m$, we will need to know that $\varphi(U)$ is open in $\mathbb{R}^m$, and here lies the problem. The solution would be trivial if we demanded that φ be surjective. The solution to the problem in general is based on a rather deep

3. Some Topological Properties

result called *invariance of domain*, which says that if U and V are homeomorphic subsets of $\mathbb{R}^m$ and U is open, then so is V open. This result, intuitively "obvious," is found, for example, on page 277 of Ref. 11 and on page 82 of Ref. 8. We will henceforth assume this result.

If (U,φ) is a coordinate chart, then φ induces a coordinate system on U, namely, φ^{-1} assigns coordinates to points of U, hence the name. This local transfer of the coordinates of $\mathbb{R}^m$ to a manifold permits the transfer of local processes which may be carried out in $\mathbb{R}^m$ to the manifold. The simplest example of this would be continuity, although, of course, we already know what is meant by continuity of a function between manifolds. Another example of a local process in $\mathbb{R}^m$ is differentiation, and in Chap. 5 we will see how to transfer this to a manifold. Of course two different coordinate charts might not induce the same coordinate systems on their overlap. Again in Chap. 5 we will see how to overcome this problem.

Examples $\mathbb{R}^m$ is an m-manifold, $\{(\mathbb{R}^m,1)\}$ being an atlas where 1 is the identity map. S^m is an m-manifold. Let $U = S^m - \{(0,\ldots,0,1)\}$ and let $V = S^m - \{(0,\ldots,0,-1)\}$. Then U and V are open subsets of S^m whose union is S^m. Define homeomorphisms $\varphi : U \to \mathbb{R}^m$, $\psi : V \to \mathbb{R}^m$ by

$$\varphi(x_0,\ldots,x_m) = \left(\frac{x_0}{1-x_m}, \ldots, \frac{x_{m-1}}{1-x_m}\right)$$

$$\psi(x,\ldots,x_m) = \left(\frac{x_0}{1+x_m}, \ldots, \frac{x_{m-1}}{1+x_m}\right)$$

Then $\{(U,\varphi),(V,\psi)\}$ is an atlas. The functions φ and ψ are called *stereographic projection* from $(0,\ldots,0,1)$ and $(0,\ldots,0,-1)$. Given the point $(x_0,\ldots,x_m) \in U$, construct the straight line joining $(0,\ldots,0,1)$ and $(x_0,\ldots,x_m)$. It (or its extension) meets $\mathbb{R}^m$ at the point $\varphi(x_0,\ldots,x_m)$, as shown in Fig. 11.

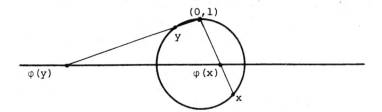

FIGURE 11

$$T^2 = \{(x,y,z) \in \mathbb{R}^3 \mid (\sqrt{x^2 + y^2} - 2)^2 + z^2 = 1\}$$

the *torus*, is a 2-manifold. The torus is obtained by revolving the circle in the xz plane centered at (2,0,0) of radius 1 about the z axis.

The torus may be made from a square sheet of rubber by gluing opposite pairs of sides. If we glue one pair of opposite sides we obtain a cylinder. The torus is obtained by bending this cylinder around until the two ends abut sufficiently to enable the gluing to take place. Reversing the order of the gluing does not change the type of intermediate object; it is still a cylinder. More importantly, it should be noted that when a pair of sides is glued together, we choose directions on each of the sides, matching the directions before gluing. For each pair, the chosen directions are the same (or parallel). This is illustrated by Fig. 12 in which the arrows show these directions.

Now suppose that we glue the first pair of sides as before so as to obtain a cylinder and then glue the second pair in such a way that the chosen direction for one side is opposite that for the other side. The resulting surface is called the *Klein bottle*, K^2. It, too, is a 2-manifold but is only one sided. This latter property can be realized easily if we reverse the order of gluing, in which case the intermediate state is the one-sided Möbius strip. Figure 13 shows the construction of K^2, but note that the picture of K^2 has a serious shortcoming: the circle of self-intersection

3. *Some Topological Properties*

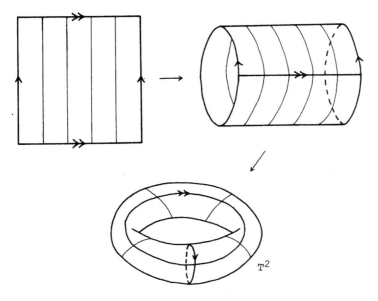

FIGURE 12

should not be there. We cannot draw a better picture as K^2 does not embed in $\mathbb{R}^3$. In $\mathbb{R}^4$ the circle of self-intersection can be eliminated very easily.

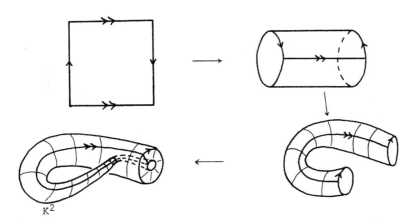

FIGURE 13

EXERCISES

1. Prove that the cofinite topology on an infinite set is compact but not Hausdorff.
2. Let X be a finite set. Prove that the only Hausdorff topology on X is the discrete topology. (Hint: Use Theorem 6 with X discrete and Y some other topology on the same set.)
3. Determine which of the following subsets of $\mathbb{R}$ or $\mathbb{R}^2$ are compact. Justify your answers.
   (a) Subsets of $\mathbb{R}$: ϕ; $\mathbb{R}$; $\{x \in \mathbb{Q} \mid 0 \leq x \leq 1\}$; $\{x \in \mathbb{Q} \mid 0 < x < 1\}$.
   (b) Subsets of $\mathbb{R}^2$: $\{(x_1,x_2) \mid 0 < x_1 x_2 < 1\}$; $\{(x_1,x_2) \mid 0 \leq x_1 x_2 \leq 1\}$; $\{(x_1,x_2) \mid x_1 = 0$ and $-1 \leq x_2 \leq 1\} \cup \{(x_1,x_2) \mid x_1 > 0$ and $x_2 = \sin(1/x_1)\}$; $\{(0,0)\} \cup \{(x_1,x_2) \mid x_1 > 0$ and $x_2 = \sin(1/x_1)\}$.
4. Determine all compact subsets of (i) the discrete space; (ii) the concrete space.
5. Check the details of the proof of Theorem 3.
6. Give an example of a space and a compact subset of this space which is not closed.
7. Verify the Heine-Borel theorem for $\mathbb{R}$: a subset of $\mathbb{R}$ is compact iff it is closed and bounded. (By Theorem 5 it is enough to show that compact sets are bounded and that closed intervals are compact. To verify the former, show instead that unbounded sets are noncompact. The latter follows from the topological invariance of compactness and the compactness of $[0,1]$.)
8. Let M^m be a manifold. Prove that M is connected iff for all $x,y \in M$, there is an open neighborhood U of both x and y such that U is homeomorphic to $\mathbb{R}^m$. (Hint: Show that for fixed $x \in M$, $\{y \in M \mid \exists$ open neighborhood U of both x and y such that U is homeomorphic to $\mathbb{R}\}$ is both open and closed in M; then appeal to Corollary 2.5.)
9. Let M be a manifold, let $x \in M$, and let N be a neighborhood of x in M. Prove that N contains a connected neighborhood of x.

10. Let X be a set and suppose we have a collection $\{(U_\alpha, \varphi_\alpha) \mid \alpha \in A\}$, where for all α, $U_\alpha \subset X$ and $\varphi_\alpha : U_\alpha \to \mathbb{R}^m$ is an injection such that $\varphi_\alpha(U_\alpha)$ is an open subset of $\mathbb{R}^m$. Suppose further that $X = \bigcup_\alpha U_\alpha$ and that for all $\alpha, \beta \in A$

$$\varphi_\beta \circ \varphi_\alpha^{-1} : \varphi_\alpha(U_\alpha \cap U_\beta) \to \mathbb{R}^m$$

is an embedding having open domain. Prove that

$$\{U \subset X \mid \exists\, \alpha \text{ with } U \subset U_\alpha \text{ and } \varphi_\alpha(U) \text{ is open in } \mathbb{R}^m\}$$

is a basis for a topology on X. Under what circumstance is this topology Hausdorff; hence X is a manifold? (Appeal to invariance of domain.)

4
SOME ADVANCED CALCULUS

In this chapter we gather together some tools from advanced calculus. In particular, we consider differentiability of a function with domain and range subsets of euclidean space, and the associated jacobian matrix. These concepts and the related results are gathered here for easy reference later. In Chap. 5, for example, we will see how the local phenomenon of differentiability is transferred to a manifold, which is locally like euclidean space. The first lemma, when transferred to a manifold, will be found to be extremely useful for the extension of functions over a manifold. The inverse function theorem and its corollaries will also be used many times in the sequel, but the result of this chapter which has the most striking consequences in our classification of surfaces is Theorem 5. The proof of this theorem is a variation of the Gram-Schmidt orthogonalization process. Familiarity with the relevant linear algebra is assumed, in particular, the rank of a matrix and determinants.

Definition Let U be an open subset of $\mathbb{R}^m$ for some m and let $f : U \to \mathbb{R}$ be a function. Say that f is *differentiable of class* C^r (where r = 1, 2, ...) if all partial derivatives of f of orders up to r exist and are continuous. Say that f is *differentiable of class* C^∞ iff f is differentiable of class C^r for all r = 1, 2, ... If A is any subset of $\mathbb{R}^m$ and $f : A \to \mathbb{R}$, then f is differentiable of class C^r (r = 1, 2, ..., ∞) iff f extends to a function whose domain is an open set containing A and which is differentiable of class C^r.

Now suppose $A \subset \mathbb{R}^m$ and $f : A \to \mathbb{R}^n$. Then f may be split into *component* or *coordinate functions*: there are n functions $f_i : A \to \mathbb{R}$ (i = 1, ..., n) such that for all $x \in A$, we have

$$f(x) = (f_1(x), \ldots, f_n(x))$$

By saying that f is differentiable of class C^r we will mean that each of the coordinate functions f_i is.

It is usual to abbreviate the above term, and any of the following expressions may be used when no confusion is likely to arise: C^r, differentiable, smooth.

Examples Any polynomial function in m variables is C^∞ on any subset of $\mathbb{R}^m$. The function $f : \mathbb{R}^m \to \mathbb{R}$ defined by $f(x) = (1 - |x|^2)^{4/3}$ is C^1 on any subset of $\mathbb{R}^m$ and is C^r ($r \geq 2$) on a subset A of $\mathbb{R}^m$ provided $A \cap S^{m-1} = \phi$; partial derivatives of f of order ≥ 2 involve $1 - |x|^2$ in the denominator.

LEMMA 1. There is a C^∞ function $h : \mathbb{R}^m \to \mathbb{R}$ satisfying:

i. $h(\mathbb{R}^m) = [0,1]$.
ii. $h^{-1}(1) = \frac{1}{2} B^m$.
iii. $h^{-1}(0) = \mathbb{R}^m - \text{Int } B^m$.

Proof: In the case m = 1, we want a C^∞ function having graph as shown in Fig. 14. The graph displays unusual, but similar,

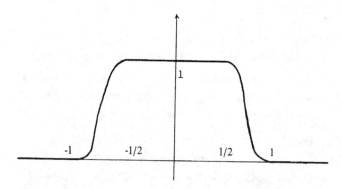

FIGURE 14

4. Some Advanced Calculus

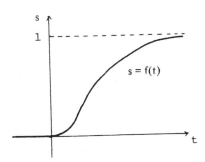
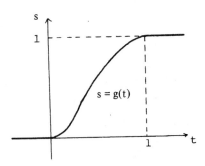

FIGURE 15

behavior at the four points $(\pm 1, 0)$, $(\pm\frac{1}{2}, 1)$: it is constant to one side of the point and nonconstant to the other. We begin the proof by displaying a function which possesses this behavior at one point. Define $f, g : \mathbb{R} \to \mathbb{R}$, whose graphs are shown in Fig. 15, as follows:

$$f(t) = \begin{cases} e^{-1/t} & \text{if } t > 0 \\ 0 & \text{if } t \leq 0 \end{cases} \qquad g(t) = \frac{f(t)}{f(t) + f(1-t)}$$

Then f, and hence g, is C^∞. Define the required h by

$$h(x) = g(\tfrac{4}{3}(1 - |x|^2))$$

The function h is essentially obtained by sliding g to the left a bit and then revolving about the axis. The number $\tfrac{4}{3}$ is there because $1 - (\tfrac{1}{2})^2$ is $\tfrac{3}{4}$. □

Definitions Let $f = (f_1, \ldots, f_n) : A \to \mathbb{R}^n$ be C^1, where $A \subset \mathbb{R}^m$. The *jacobian matrix* of f at x is the matrix $Df(x)$ whose (i,j) entry is the value of $\partial f_i / \partial x_j$ at x. Note that the matrix $Df(x)$ is of size $n \times m$. In Int A each of the entries of $Df(x)$ depends continuously on x, so Df is effectively a continuous function from Int A to $\mathbb{R}^{mn}$.

The rank of the jacobian matrix $Df(x)$ is called the *rank* of f at x. If $m = n$, the determinant of the jacobian matrix is called the *jacobian determinant* and is denoted $\Delta f(x)$.

Let $f : U \to V$, where U and V are open subsets of $\mathbb{R}^m$. Then f is called a C^r *diffeomorphism* iff f is a homeomorphism and each of f and f^{-1} is C^r.

Note that if f is a diffeomorphism, then its jacobian matrix is nonsingular: in fact $[Df(x)]^{-1} = Df^{-1}(f(x))$, this following from the *chain rule*. The chain rule says that if $f : U \to \mathbb{R}^n$ and $g : V \to \mathbb{R}^p$ are C^r at $x \in U$ and $f(x) \in V$, respectively, where U is open in $\mathbb{R}^m$ and V is open in $\mathbb{R}^n$, then gf is C^r at x and

$$D(gf)(x) = Dg(f(x))Df(x)$$

where the multiplication on the right is matrix multiplication.

Examples The identity function is a diffeomorphism. The homeomorphism $f : \mathbb{R} \to \mathbb{R}$ defined by $f(x) = x^3$ is not a C^r diffeomorphism for any r even though f is C^∞; f^{-1} is not C^1 at 0. The function $g : (-\pi/2, \pi/2) \to \mathbb{R}$ defined by $g(t) = \tan t$ is a C^∞ diffeomorphism. The function $h : \mathbb{R} \to \mathbb{R}$ defined by $h(x) = 2x + x|x|$ is a C^1 diffeomorphism but not a C^r diffeomorphism for $r > 1$.

THEOREM 2 (Inverse Function Theorem). Let U be an open subset of $\mathbb{R}^m$ and let $f : U \to \mathbb{R}^m$ be a C^r function. Let $x_0 \in U$ and suppose that $Df(x_0)$ is nonsingular. Then f is a C^r diffeomorphism of some neighborhood of x_0 onto some neighborhood of $f(x_0)$.

Proof: Omitted. See Appendix A. □

The jacobian matrix of a function at a point gives a linear approximation to that function at the point. The inverse function theorem carries this one step further; if the jacobian matrix is nonsingular (so that domain and range must have the same dimension) then the linear approximation has an inverse which is itself the linear approximation of the inverse of the function.

4. Some Advanced Calculus

Sometimes one can use the inverse function theorem to deduce that a function is a (global) diffeomorphism.

Example Let $g : (-\pi/2, \pi/2) \to \mathbb{R}$ be the C^∞ homeomorphism given by $g(t) = \tan t$. Since $g'(t) = \sec^2 t \neq 0$ for all $t \in (-\pi/2, \pi/2)$, by the inverse function theorem, g^{-1} is also C^∞. Thus g is a diffeomorphism.

COROLLARY 3. Let U be an open neighborhood of 0 in $\mathbb{R}^m$ and let $f : U \to \mathbb{R}^n$ be a C^r function having rank m at 0 (so $m \leq n$). Suppose $f(0) = 0$. Then there is a C^r diffeomorphism g of a neighborhood of 0 in $\mathbb{R}^n$ onto another such neighborhood such that

$$gf(x_1, \ldots, x_m) = (x_1, \ldots, x_m, 0, \ldots, 0)$$

Proof: Case I. Suppose that the $m \times m$ matrix with (i,j) entry $\partial f_i / \partial x_j$ is nonsingular at 0, i.e., the first m rows of $Df(0)$ are linearly independent. Define

$$F : U \times \mathbb{R}^{n-m} \to \mathbb{R}^n$$

by $F(x_1, \ldots, x_n) = f(x_1, \ldots, x_m) + (0, \ldots, 0, x_{m+1}, \ldots, x_n)$. Then $DF(x)$ has the form

$$\left(\begin{array}{c|c} Df(x) & 0 \\ \hline & \text{Identity} \end{array} \right)$$

so is nonsingular at 0. The local inverse, say g, of F given by the inverse function theorem satisfies the required conditions.

Case II (General case). Exercise: Some m rows of $Df(0)$ are linearly independent, so the definition of F should be modified so that the $n - m$ 1's in the last columns of $DF(x)$ go into the other rows. □

COROLLARY 4. Let U be an open neighborhood of 0 in $\mathbb{R}^m$ and let $f : U \to \mathbb{R}^n$ be a C^r function having rank n at 0 (so $m \geq n$).

Suppose $f(0) = 0$. Then there is a C^r diffeomorphism h of a neighborhood of 0 in $\mathbb{R}^m$ onto another such neighborhood such that $h(0) = 0$ and

$$fh(x_1,\ldots,x_m) = (x_1,\ldots,x_n)$$

Proof: Exercise. □

We now restrict our attention to the case where the range has dimension 1.

Definition Let $f : \mathbb{R}^m \rightarrow \mathbb{R}$ be C^r ($r \geq 2$). A point $x \in \mathbb{R}^m$ is a *critical point* of f iff all entries of the jacobian matrix $Df(x)$ are 0. The *hessian* of f at x is the $m \times m$ matrix $Hf(x)$ whose (i,j) entry is $\partial^2 f/\partial x_i \partial x_j\big|_x$. A critical point x of f is *(non-)degenerate* iff $Hf(x)$ is (non-)singular.

Examples $f : \mathbb{R} \rightarrow \mathbb{R}$ defined by $f(x) = x^2 + 2x + 1$ has a nondegenerate critical point at $x = -1$; $g : \mathbb{R} \rightarrow \mathbb{R}$ defined by $g(x) = x^4$ has a degenerate critical point at $x = 0$; $h : \mathbb{R}^2 \rightarrow \mathbb{R}$ defined by $h(x,y) = x^2 + 4xy + 4y^2$ has a degenerate critical point at $(x,y) = (0,0)$. In fact, the graph of h in $\mathbb{R}^3$ is a parabolic cylinder tangent to the xy plane on the line $x + 2y = 0$; every point on this line is critical. If we define $i : \mathbb{R}^2 \rightarrow \mathbb{R}$ by $i(x,y) = y^3 - 3x^2 y$, then $(0,0)$ is a degenerate critical point of i, even though $(0,0)$ is the only critical point. The graph of i is known as a monkey saddle; three ridges and three valleys lead from the saddle point.

The following result is probably the most useful result from this chapter in our application to the classification of surfaces. Surfaces will be classified by studying smooth functions from them to $\mathbb{R}$. By a geographical analogy, the inverse image under such a function of a point in $\mathbb{R}$ will be called a level. It will be found that changes in a level occur only at critical points. These changes can be classified by use of the standard forms of such

4. Some Advanced Calculus

functions about critical points as given by the following result. The proof of the result involves an extension of the Gram-Schmidt diagonalization process of linear algebra.

THEOREM 5. Let 0 be a nondegenerate critical point of the C^r ($r \geq 2$) function $f : \mathbb{R}^m \to \mathbb{R}$. Then there is a diffeomorphism θ of a neighborhood of 0 in $\mathbb{R}^m$ onto another such neighborhood with $\theta(0) = 0$, and there are numbers $c_i = \pm 1$ ($i = 1, \ldots, m$) such that for all $z = (z_i)$ in the domain of θ,

$$f\theta(z) = \sum_{i=1}^{m} c_i z_i^2 + f(0)$$

Proof: The proof consists of two steps. In the first step, we obtain a variation of Taylor's theorem. This is then subjected, in the second step, to the diagonalization process.

Step I. For all $x \in \mathbb{R}^m$, we have

$$f(x) - f(0) = \int_0^1 \frac{d}{dt} f(tx)\, dt \qquad \text{(fundamental theorem of calculus)}$$

$$= \int_0^1 \sum_{i=1}^{m} x_i \left.\frac{\partial f}{\partial x_i}\right|_{tx} dt \qquad \text{(chain rule)}$$

$$= \sum_{i=1}^{m} x_i \int_0^1 \left.\frac{\partial f}{\partial x_i}\right|_{tx} dt$$

Let $f_i(x) = \int_0^1 \left.\partial f/\partial x_i\right|_{tx} dt$. Then we have

$$f(x) = \sum_{i=1}^{m} x_i f_i(x) + f(0)$$

Now let $f_{ij}(x) = \int_0^1 \left.\partial f_i/\partial x_j\right|_{tx} dt$. Since 0 is a critical point, $\left.\partial f/\partial x_i\right|_0 = 0$, so that $f_i(0) = 0$ and hence

$$f_i(x) = \sum_{j=1}^{m} x_j \int_0^1 \partial f_i/\partial x_j \big|_{tx} \, dt \, . \quad \text{Thus}$$

$$f(x) = \sum_{i=1}^{m} \sum_{j=1}^{n} x_i x_j f_{ij}(x) + f(0)$$

Note that we have expressed f in the form

$$f(x) = x(f_{ij}(x)) \, x^* + f(0)$$

where x is the matrix $(x_1, \ldots, x_m)$ and x* is its transpose. The m × m matrix $(f_{ij}(x))$ is symmetric with rank m at 0 (by nondegeneracy) and hence, by continuity of the f_{ij}, the rank is m in a neighborhood of 0 . We are called on to diagonalize this matrix in such a way that the elements in the diagonal are ±1 .

Step II. Recall the following fact from linear algebra. If A is a real symmetric m × m matrix, then there is an orthogonal matrix T such that T*AT is diagonal, the diagonal elements being the eigenvalues of A . Again, T* denotes the transpose of T , and orthogonality means that T*T = TT* = 1 . One can verify that if the entries in the matrix A are allowed to vary differentiably by a small amount, then the corresponding matrix T also varies differentiably. Applying this to the matrix $(f_{ij}(x))$, we obtain an orthogonal matrix T(x) , defined for x in a neighborhood of 0 in $\mathbb{R}^m$, such that

$$T(x)*(f_{ij}(x))T(x) = \text{diag}(\lambda_1(x), \ldots, \lambda_m(x))$$

where $\lambda_1(x), \ldots, \lambda_m(x)$ are the (differentiably varying) eigenvalues of the matrix $(f_{ij}(x))$.

Define the function g in a neighborhood of 0 in $\mathbb{R}^m$ to $\mathbb{R}^m$ by g(x) = xT(x) . Then Dg(0) = T(0) is nonsingular, so by the inverse function theorem, g is a diffeomorphism between two neighborhoods of 0 in $\mathbb{R}^m$. Suppose y is an element of the image of g , say y = g(x) . Thus y = xT(x) , so that x = yT(x)* and x* = T(x)y* .

4. Some Advanced Calculus

Hence

$$\begin{aligned}
fg^{-1}(y) &= f(x) \\
&= x[f_{ij}(x)]x^* + f(0) \\
&= y\, T(x)^*[f_{ij}(x)]T(x)y^* + f(0) \\
&= y\, \text{diag}(\lambda_1(x),\ldots,\lambda_m(x))y^* + f(0) \\
&= \sum_{i=1}^{m} \lambda_i(x) y_i^2 + f(0)
\end{aligned}$$

Note that nondegeneracy of the critical point means that the eigenvalues $\lambda_1(0), \ldots, \lambda_m(0)$ are all nonzero, so $\lambda_1(x), \ldots, \lambda_m(x)$ are all nonzero in a neighborhood of 0. Assume that the domain of g is small enough so that this is the case. Define h on the image of g to $\mathbb{R}^m$ by

$$h(y) = (\sqrt{|\lambda_i g^{-1}(y)|}\; y_i) = (\sqrt{|\lambda_1(x)|}\; y_1, \ldots, \sqrt{|\lambda_m(x)|}\; y_m)$$

By the inverse function theorem, h is also a diffeomorphism in a neighborhood of 0. Let the required diffeomorphism θ be the inverse of hg, i.e., $\theta = g^{-1}h^{-1}$. Let $c_i = -1$ if $\lambda_i(0) < 0$ and $c_i = 1$ if $\lambda_i(0) > 0$. Note that if we are considering small enough neighborhoods of 0 then $\lambda_i(0) < 0$ iff $\lambda_i(x) < 0$ throughout the neighborhood.

Suppose that $z = (z_i)$ is in the domain of θ. Let $y = h^{-1}(z)$ and $x = g^{-1}(y)$. Then

$$\begin{aligned}
f\theta(z) &= fg^{-1}h^{-1}(z) \\
&= fg^{-1}(y) \\
&= \sum_{i=1}^{m} \lambda_i(x) y_i^2 + f(0) \\
&= \sum_{i=1}^{m} c_i |\lambda_i(x)| y_i^2 + f(0) \\
&= \sum c_i [\sqrt{|\lambda_i(x)|}\; y_i]^2 + f(0) \\
&= \sum c_i z_i^2 + f(0) \quad \square
\end{aligned}$$

Definition Let $f : \mathbb{R}^m \to \mathbb{R}$ be differentiable with a nondegenerate critical point at 0. Then the number of integers i for which $c_i = -1$ in Theorem 5 is called the *index* of the critical point.

As in Sylvester's law of inertia, the index is well defined. In fact it is the number of negative eigenvalues of the matrix $(f_{ij}(0))$. Since we have shown that

$$f(x) = \sum_{i=1}^{m} \sum_{j=1}^{m} x_i x_j f_{ij}(x) + f(0)$$

on twice differentiating, we find that the (i,j) entry of the hessian $Hf(0)$ is $f_{ij}(0) + f_{ji}(0) = 2f_{ij}(0)$. Thus we have $Hf(0) = 2(f_{ij}(0))$, so that λ is an eigenvalue of $(f_{ij}(0))$ iff 2λ is an eigenvalue of $Hf(0)$. In particular, $Hf(0)$ has the same number of negative eigenvalues as has $(f_{ij}(0))$, so to calculate the index, it suffices to determine the number of negative eigenvalues of the hessian.

Example Let $f(x,y) = \tan xy$. Then

$$Df(x,y) = (y \sec^2 xy \quad x \sec^2 xy)$$

$$Hf(x,y) = \begin{pmatrix} 2y^2 \sec^2 xy \tan xy & \sec^2 xy + 2xy \sec^2 xy \tan xy \\ \sec^2 xy + 2xy \sec^2 xy \tan xy & 2x^2 \sec^2 xy \tan xy \end{pmatrix}$$

Thus $(0,0)$ is a nondegenerate critical point since

$$Df(0,0) = (0,0) \qquad Hf(0,0) = \begin{pmatrix} 0 & 1 \\ 1 & 0 \end{pmatrix}$$

$$\det(\lambda 1 - Hf(0,0)) = \det \begin{pmatrix} \lambda & -1 \\ -1 & \lambda \end{pmatrix} = \lambda^2 - 1 = (\lambda - 1)(\lambda + 1)$$

Since $Hf(0,0)$ has one negative eigenvalue, the index is 1.

Critical points of index 0 are local minima and critical points of index m are local maxima. If $m = 2$, a critical point of index 1

4. *Some Advanced Calculus*   49

is a saddle. With appropriate modifications, the above discussion and theorem holds for critical points other than 0.

Note the connection between Theorem 5 and the standard second derivative test for critical points of a function from $\mathbb{R}^2$ to $\mathbb{R}$.

We complete this chapter with a lemma which, although not directly used, serves as a model for other, unstated, lemmas which are used.

LEMMA 6   (Prototype). Let $f : \mathbb{R}^m \to \mathbb{R}$ be defined by

$$f(x_1,\ldots,x_m) = -\sum_{i=1}^{\lambda} x_i^2 + \sum_{i=\lambda+1}^{m} x_i^2$$

Then for all $\alpha > 0$, there is a C^∞ function $g : \mathbb{R}^m \to \mathbb{R}$ satisfying

i. $g(x) = f(x)$ for all $x \in \mathbb{R}^m - (\alpha B^\lambda \times 2B^{m-\lambda})$.

ii. $g(0) = 1$.

iii. 0 is the only critical point of g, is nondegenerate, and has index λ.

Fig. 16 illustrates this lemma in the case where $\lambda = 1$ and $m = 2$. The saddle point is raised one unit while the amount by which points on the graph is raised tapers off to zero as we move off to the edge of the rectangle $[-\alpha,\alpha] \times [-2,2]$.

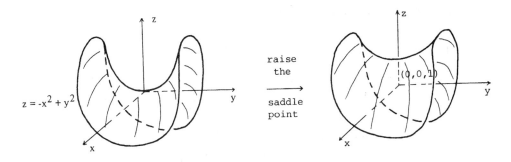

FIGURE 16

Proof: Given $\alpha > 0$, as in Lemma 1 construct a C^∞ function $h : \mathbb{R}^m \to \mathbb{R}$ such that

$$h(\mathbb{R}^m) = [0,1]$$

$h^{-1}(1)$ is a neighborhood of 0, say $\left(\frac{\alpha}{2} B^\lambda\right) \times B^{m-\lambda}$

$$h(\mathbb{R}^m - (\alpha B^\lambda \times 2B^{m-\lambda})) = 0$$

$$x_i \frac{\partial h}{\partial x_i} \leq 0 \text{ if } i \leq \lambda$$

for $i > \lambda$, $2x_i + \frac{\partial h}{\partial x_i} \begin{cases} > 0 & \text{if } x_i > 0 \\ < 0 & \text{if } x_i < 0 \end{cases}$

Note that h must decrease between $(\alpha/2) B^\lambda \times B^{m-\lambda}$ and $\mathbb{R}^m - (\alpha B^\lambda \times 2B^{m-\lambda})$. The last condition says that this decrease is not too rapid.

Let $g = f + h$. Then g is C^∞, clearly satisfying i and ii. Further, $\partial g/\partial x_i = \partial f/\partial x_i + \partial h/\partial x_i = \pm 2x_i + \partial h/\partial x_i$. For $i \leq \lambda$, $\partial g/\partial x_i = -2x_i + \partial h/\partial x_i = 0$ iff $x_i = 0$, since $\partial h/\partial x_i$ has the same sign as $-2x_i$. For $i > \lambda$, $\partial g/\partial x_i = 2x_i + \partial h/\partial x_i = 0$ iff $x_i = 0$ by the last condition on h. Thus the only critical point of g is 0. Since h is constant in a neighborhood of 0, the critical point must be nondegenerate of index λ. □

EXERCISES

1. Determine the rank of the function $f : \mathbb{R}^2 \to \mathbb{R}^2$ defined by $f(x,y) = (x^2 + 2xy + y^2, xy^2 + x^2y)$ at each point of its domain. Is f a diffeomorphism?
2. The transformation $(r,\theta,\varphi) \to (r \cos\theta \sin\varphi, r \sin\theta \sin\varphi, r \cos\varphi)$, $0 \leq r < \infty$, $0 \leq \theta < 2\pi$, $0 \leq \varphi \leq \pi$, transforms the spherical polar coordinates of a point in $\mathbb{R}^3$ into its rectangular cartesian coordinates (Fig. 17). Determine the rank of this function at each point of its domain. Hence, by

4. Some Advanced Calculus   51

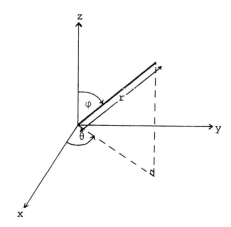

FIGURE 17

geometrically interpreting the above, show that
$\mathbb{R}^3 - \{(x,y,z) \in \mathbb{R}^3 \mid y = 0 \text{ and } x \geq 0\}$ is diffeomorphic to $\mathbb{R}^3$.

3. The function $f : \mathbb{R}^2 \to \mathbb{R}^3$ defined by

$$f(x,y) = (x + 2y, x + y, xy + 2x + 3y)$$

satisfies the hypotheses of Corollary 3. Find a diffeomorphism g whose existence is assured by Corollary 3.

4. Prove Corollary 4.

5. For each of the following functions, verify that 0 is a critical point, determine whether the critical point is nondegenerate, and if so, determine the index.

$f : \mathbb{R}^2 \to \mathbb{R}$ defined by $f(x,y) = 5x^2 - 2xy + 2y^2$
$g : \mathbb{R}^2 \to \mathbb{R}$ defined by $g(x,y) = 5xy - 3x^2$
$h : \mathbb{R}^2 \to \mathbb{R}$ defined by $h(x,y) = \sin(\sin xy)$
$i : \mathbb{R}^3 \to \mathbb{R}$ defined by $i(x,y,z) = xy + yz + zx$
$j : \mathbb{R}^3 \to \mathbb{R}$ defined by $j(x,y,z) = x^2 + y^2 + z^2 + xy + yz + zx$

6. Let $f : \mathbb{R}^m \to \mathbb{R}$ be a differentiable function having a nondegenerate critical point at p, and let $h : \mathbb{R}^m \to \mathbb{R}^m$ be a diffeomorphism. Let $q = h^{-1}(p)$. Prove that $fh : \mathbb{R}^m \to \mathbb{R}$

has a nondegenerate critical point at q. [Hints: Using Theorem 5, we may reduce consideration to the case where $p = 0$ and $f(x) = \sum_{i=1}^{m} c_i x_i^2$ ($c_i = \pm 1$), since having dealt with this case, then letting θ be a diffeomorphism of a neighborhood of 0 onto a neighborhood of p as given by Theorem 5, we have $fh = (f\theta)(\theta^{-1}h)$. Applying the special case to $f\theta$ and $\theta^{-1}h$ gives us the general case. Now for the special case: show by differentiation that

$$H(fh)(q) = D(h)(q)^* \times H(f)(0) \times D(h)(q)$$

a product of three matrices each of which is nonsingular. In fact, keeping in mind that p is a critical point, this equation holds in the general case.]

5
DIFFERENTIABLE MANIFOLDS

In this chapter we show how to combine the calculus of Chap. 4 with the topological manifolds of Chap. 3. We see how to decide whether a given function between two manifolds is differentiable. The essential features are that differentiation is a local process and that a manifold is locally like the arena of calculus, viz., $\mathbb{R}^m$. Charts are chosen, one about a particular point in the domain of the function and one about the image point, and these are used to transfer the question of differentiability at the point back to euclidean space. To ensure that the differentiability of a function is independent of the choices of charts, we must restrict the collection of charts from which the selection is made. This gives rise to the notion of a differentiable manifold. Once we have determined how to answer the question of differentiability, we will see what other concepts from Chap. 4 generalize to the context of a manifold. The jacobian matrix depends on the charts chosen, so is lost. However, the notion of rank is independent of charts chosen, so it carries over to a manifold, where it is an important concept. Two other notions which carry over to a manifold are critical point and nondegeneracy, but a study of these is deferred until Chap. 9.

Definition Let M^m be a topological manifold. By a *differential structure of class* C^r or a C^r *structure* on M, we mean an atlas $\mathcal{D}$ satisfying:

DS1. For all $(U,\varphi),(V,\psi) \in \mathcal{D}$ the coordinate transformation

$$\psi\varphi^{-1} : \varphi(U \cap V) \longrightarrow \mathbb{R}^m$$

is of class C^r.

DS2. $\mathcal{D}$ is maximal with respect to DS1, i.e., if (W,X) is a chart of M which is not in $\mathcal{D}$, then $\mathcal{D} \cup \{(W,X)\}$ does not satisfy DS1 [i.e., there exists $(U,\varphi) \in \mathcal{D}$ such that at least one of the coordinate transformations φX^{-1}, $X\varphi^{-1}$ is not C^r].

If $\mathcal{B}$ is an atlas on M satisfying only DS1, then it is routine to show that there is a unique differential structure $\mathcal{D}$ of class C^r on M such that $\mathcal{B} \subset \mathcal{D}$. $\mathcal{B}$ is called a *basis* for $\mathcal{D}$.

A *differentiable manifold of class* C^r or a C^r *manifold* is a pair $(M,\mathcal{D})$, where M is a topological manifold and $\mathcal{D}$ is a differential structure of class C^r on M. Henceforth when we speak of a manifold we will mean a differentiable manifold, often omitting reference to the associated structure.

The definition above is probably quite formidable at first sight, so we will spend some time discussing it. We begin by discussing DS1. For many purposes it is enough to consider only a basis for a differential structure; cf. the similar situation regarding topological spaces. This is the case, for example, when we consider the notion of differentiability of a function; see Corollary 3. Suppose, then, that we have a basis $\mathcal{B}$ for a differential structure. Note that DS1 tells us that each coordinate transformation is a diffeomorphism, since both $\psi\varphi^{-1}$ and $(\psi\varphi^{-1})^{-1} = \varphi\psi^{-1}$ are differentiable.

A chart may be used to impose on part of a manifold the coordinate system of $\mathbb{R}^m$. The coordinate systems imposed by neighboring charts might not match up precisely. This would be too much to expect; cf. Fig. 18. DS1 says that at least there is a diffeomorphism throwing one onto the other. The significance of this will become more apparent when we define what it means for a function to be differentiable.

5. Differentiable Manifolds

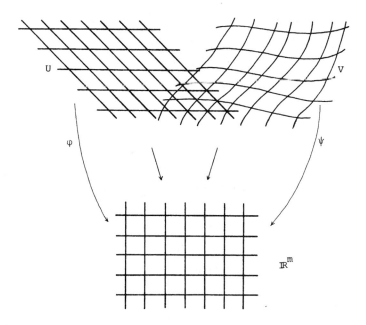

FIGURE 18

Examples $\{(\mathbb{R}^m, 1)\}$ is a basis for a differential structure of class C^∞ on $\mathbb{R}^m$. It is clearly an atlas, and one can trivially verify the condition DS1 [trivially, since DS1 need only be checked in the case $(U,\varphi) = (V,\psi) = (\mathbb{R}^m, 1)$]. This is the *natural* or *usual* differential structure on $\mathbb{R}^m$.

Any homeomorphism $h : \mathbb{R}^m \longrightarrow \mathbb{R}^m$ gives a basis for a differential structure on $\mathbb{R}^m$, viz., $\{(\mathbb{R}^m, h)\}$. Again it is trivial to check DS1.

Let (U,φ) and (V,ψ) be the stereographic projection charts of Chap. 3. As noted there, $\{(U,\varphi),(V,\psi)\}$ is an atlas on S^m. One can verify that

$$\varphi^{-1}(y_1,\ldots,y_m) = \left(\frac{2y_1}{\sum_{i=1}^m y_i^2 + 1}, \ldots, \frac{2y_m}{\sum_{i=1}^m y_i^2 + 1}, \frac{\sum_{i=1}^m y_i^2 - 1}{\sum_{i=1}^m y_i^2 + 1} \right)$$

Thus we have

$$\psi\varphi^{-1}(y_1,\ldots,y_m) = \left(\frac{y_1}{\sum_{i=1}^m y_i^2},\ldots,\frac{y_m}{\sum_{i=1}^m y_i^2}\right)$$

This formula is valid on $\varphi(U \cap V) = \mathbb{R}^m - \{0\}$, on which $\sum_{i=1}^m y_i^2 \neq 0$. Since each coordinate function is a rational function, it is C^∞, so that $\psi\varphi^{-1}$ is C^∞. By symmetry $\varphi\psi^{-1}$ is C^∞ and hence the atlas $\{(U,\varphi),(V,\psi)\}$ is a basis for a differential structure of class C^r for any $r = 1, 2, \ldots, \infty$. This is the *usual* or *natural* structure on S^m.

The torus and Klein bottle can also be given natural differential structures, the torus as follows. Note that on the torus we have natural senses of latitude and longitude. Lines of longitude are obtained by revolving a point on the circle in the xz plane centered at (2,0,0) of radius 1 about the z axis, and lines of latitude are obtained by intersecting T^2 with a half plane bounded by the z axis. In this way we can define (modulo 2π) the latitude and longitude of a point on T^2. The latitude is the angle α from the positive x axis to the half plane determining the line of latitude, and the longitude is the angle β in the xz plane from the positive x axis to the appropriate line of longitude; cf. Fig. 19.

Let U be the torus with the π line of longitude (i.e., the innermost circle) removed, and V be the torus with the 0 line of longitude (i.e., the outermost circle) removed. Then $\{U,V\}$ is an open cover of T^2. Having removed the innermost circle from T^2, we can unroll U to obtain an annulus in $\mathbb{R}^2$ (i.e., region between two concentric circles). This gives us an embedding $\varphi : U \to \mathbb{R}^2$. Similarly we can unroll V to give an embedding $\psi : V \to \mathbb{R}^2$. Provided the unrolling is done smoothly, $\{(U,\varphi),(V,\psi)\}$ will be a basis for a C^r structure on T^2, $(r = 1, 2, \ldots, \infty)$. More precisely, define $\varphi(p)$ and $\psi(p)$ to be $(\beta \cos \alpha, \beta \sin \alpha)$, where α and β are, respectively, the latitude and longitude of the point p. In the case of φ, we assume $\beta \in (\pi, 3\pi)$, and in the case of ψ,

5. Differentiable Manifolds

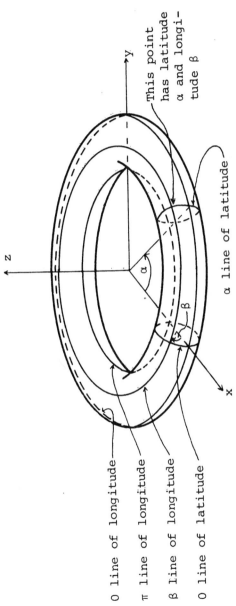

FIGURE 19

we assume $\beta \in (0, 2\pi)$. Then φ sends U onto the annulus in $\mathbb{R}^2$ between the circles of radii π and 3π centered at 0 ; ψ sends V onto the annulus in $\mathbb{R}^2$ between the circles of radii 0 and 2π centered at 0.

To verify DS1, note that $\varphi(U \cap V)$ consists of two annuli, separated by the circle of radius 2π. On the inner annulus $\psi\varphi^{-1}$ is the identity, so is C^∞. On the outer annulus, $\psi\varphi^{-1}$ moves each point radially towards the origin a distance 2π, i.e., if $2\pi < \sqrt{x^2 + y^2} < 3\pi$, then

$$\psi\varphi^{-1}(x,y) = \frac{\sqrt{x^2+y^2} - 2\pi}{\sqrt{x^2+y^2}} (x,y)$$

Thus $\psi\varphi^{-1}$ is C^∞ on the outer annulus also. By symmetry $\varphi\psi^{-1}$ is also C^∞.

We now turn to a discussion of DS2. We start with a simpler analogue. Let S be a subset of the real numbers satisfying:

i. For all $a, b \in S$, $ab > 0$.
ii. S is maximal with respect to i.

There are two possibilities for S : the positive reals and the negative reals. Any nonempty subset T of the real numbers satisfying i (hence all elements of T have the same sign) might be called a basis. For many purposes, it might be enough to consider only T, but sometimes we might require another real number not in T but having the same sign. We might then need to appeal to ii, enlarging T by use of ii. This is the kind of role played by DS2.

Given a basis B for a differential structure $\mathcal{D}$ of class C^r, we can recreate the differential structure thus:

$$\mathcal{D} = \{(V, \psi) \mid (V, \psi) \text{ is a chart for M and for all } (U, \varphi) \in B$$
$$\varphi\psi^{-1} \text{ and } \psi\varphi^{-1} \text{ are of class } C^r \text{ where defined}\}$$

Thus $\mathcal{D}$ will be a huge set (uncountable) even though B may be small — any compact manifold has a finite basis.

5. Differentiable Manifolds

Examples The basis $\{(\mathbb{R}^m, 1)\}$ gives rise to the differential structure

$$\{(U, \varphi) \mid U \text{ is open in } \mathbb{R}^m \text{ and } \varphi : U \to \mathbb{R}^m \text{ is a differentiable embedding such that } \varphi^{-1} : \varphi(U) \to \mathbb{R}^m \text{ is also differentiable}\}$$

Thus if $h : \mathbb{R}^m \to \mathbb{R}^m$ is a homeomorphism, then $(\mathbb{R}^m, h)$ is in the natural structure on $\mathbb{R}^m$ iff h is a diffeomorphism. If h is not a diffeomorphism, then $\{(\mathbb{R}^m, h)\}$ is a basis for a different differential structure on $\mathbb{R}^m$. The calculus we might carry out in this latter case will, in a sense, be unnatural.

To take a specific example, for each $r > 0$, define $h_r : \mathbb{R} \to \mathbb{R}$ as in Fig. 20 by $h_r(t) = t$ if $t \leq 0$ and $h_r(t) = rt$ if $t \geq 0$. Note that h_r is a homeomorphism, but a diffeomorphism only when $r = 1$. Let $B_r = \{(\mathbb{R}, h_r)\}$. Then B_r is a basis for a differential structure $\mathcal{D}_r$. Of course $B_r \subset \mathcal{D}_r$, but if $s \neq r$, then $(\mathbb{R}, h_s) \notin \mathcal{D}_r$, since $h_r h_s^{-1}$ is not differentiable of any class. Thus for $r \neq s$, we have $\mathcal{D}_r \neq \mathcal{D}_s$, so we have uncountably many structures. Away from 0, calculus in $(\mathbb{R}, \mathcal{D}_r)$ is the same as calculus in $(\mathbb{R}, \mathcal{D}_s)$, but at 0 the two calculi are different. For example, a particle might be moving at a constant speed with respect to the natural structure, but as far as the structure $\mathcal{D}_r$ is concerned, the speed will suddenly increase at 0 if $r < 1$, then continue at the new, higher, level.

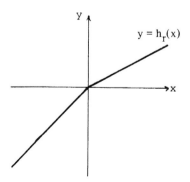

FIGURE 20

The following lemma gives us a typical use of the maximality condition DS2.

LEMMA 1. Let $(M^m, \mathcal{D})$ be a differentiable manifold and let $x \in M$. Then there exists $(U, \varphi) \in \mathcal{D}$ such that $x \in U$ and $\varphi(x) = 0$.

Proof: Since $\mathcal{D}$ is an atlas, there exists $(U, \alpha) \in \mathcal{D}$ such that $x \in U$. Let $\tau : \mathbb{R}^m \to \mathbb{R}^m$ be the translation defined by $\tau(y) = y - \alpha(x)$. Note that τ is a C^∞ diffeomorphism. Let $\varphi = \tau\alpha : U \to \mathbb{R}^m$. Then $\varphi(x) = \tau(\alpha(x)) = \alpha(x) - \alpha(x) = 0$. It is claimed that $(U, \varphi) \in \mathcal{D}$. By DS2, we must show that for all $(V, \psi) \in \mathcal{D}$, the coordinate transformations $\psi\varphi^{-1}$ and $\varphi\psi^{-1}$ are both differentiable. But

$$\psi\varphi^{-1} = \psi(\tau\alpha)^{-1} = (\psi\alpha^{-1})\tau^{-1}$$

Since $(U, \alpha), (V, \psi) \in \mathcal{D}$, $\psi\alpha^{-1}$ is differentiable, so, since τ is a diffeomorphism, $\psi\varphi^{-1}$ is differentiable. Similarly $\varphi\psi^{-1} = \tau(\alpha\psi^{-1})$ is differentiable. □

Standard mathematical procedure requires us to decide which functions preserve our new structure.

Definition Let M^m and N^n be differentiable manifolds of class at least C^r. A function $f : M \to N$ is *differentiable of class* C^r iff for every pair of charts $(U, \varphi), (V, \psi)$ in the structures of M and N, respectively, the composition

$$\psi f \varphi^{-1} : \varphi(U \cap f^{-1}(V)) \to \mathbb{R}^n$$

is of class C^r. In effect, we transfer the question of differentiability back to $\mathbb{R}^m$ and $\mathbb{R}^n$ by use of the charts.

In fact, to verify differentiability at a point, we need verify differentiability of $\psi f \varphi^{-1}$ for only a pair of charts, one about the point and one about its image, as the following lemma shows. The corollary points out that to verify differentiability of a function, we need verify differentiability of $\psi f \varphi^{-1}$ only when (U, φ) and (V, ψ) range through respective bases of M and N.

5. Differentiable Manifolds

LEMMA 2. The function $f : M^m \to N^n$ between the two differentiable manifolds $(M,\mathcal{D})$ and (N,E) is differentiable if for all $x \in M$, there exist $(U,\varphi) \in \mathcal{D}$, $(V,\psi) \in E$ such that $x \in U$, $f(x) \in V$, and $\psi f \varphi^{-1}$ is differentiable at $\varphi(x)$.

Proof: Let $(O,\alpha) \in \mathcal{D}$ and $(P,\beta) \in E$. To show that $\beta f \alpha^{-1}$ is differentiable, it suffices to verify this at x, for all $x \in \alpha(O \cap f^{-1}(P))$. Given $x \in \alpha(O \cap f^{-1}(P))$, there are charts $(U,\varphi) \in \mathcal{D}$ and $(V,\psi) \in E$ such that $\alpha^{-1}(x) \in U$, $f\alpha^{-1}(x) \in V$, and $\psi f \varphi^{-1}$ is differentiable at $\varphi \alpha^{-1}(x)$. Thus by DS1 and the chain rule,

$$\beta f \alpha^{-1} = (\beta \psi^{-1})(\psi f \varphi^{-1})(\varphi \alpha^{-1})$$

is differentiable at x. □

COROLLARY 3. Suppose $f : M^m \to N^n$ is a function between the two differentiable manifolds M and N having bases $\mathcal{B}$ and $\mathcal{C}$, respectively. Suppose that for all $(U,\varphi) \in \mathcal{B}$ and all $(V,\psi) \in \mathcal{C}$, the function $\psi f \varphi^{-1} : \varphi(U \cap f^{-1}(V)) \to \mathbb{R}^n$ is differentiable. Then f is differentiable. □

Clearly the notion of jacobian of a differentiable function in euclidean space does not easily carry over to differentiable functions between manifolds. It does, but we must reinterpret the jacobian. Surprisingly, however, rank does easily transfer. This turns out to be very important.

LEMMA 4. Let $f : M \to N$ be differentiable, let $x \in M$, and let $(U,\varphi),(O,\alpha)$ be charts in the structure of M such that $x \in U \cap O$, and $(V,\psi),(P,\beta)$ be charts in the structure of N such that $f(x) \in V \cap P$. Then the ranks of $\psi f \varphi^{-1}$ at $\varphi(x)$ and $\beta f \alpha^{-1}$ at $\alpha(x)$ are equal.

Proof: Since

$$\beta f \alpha^{-1} = (\beta \psi^{-1})(\psi f \varphi^{-1})(\varphi \alpha^{-1})$$

$$D(\beta f \alpha^{-1})(\alpha(x)) = D(\beta \psi^{-1})(\psi f(x)) \times D(\psi f \varphi^{-1})(\varphi(x)) \times D(\varphi \alpha^{-1})(\alpha(x))$$

Since $\varphi\alpha^{-1}$ and $\beta\psi^{-1}$ are diffeomorphisms, $D(\varphi\alpha^{-1})(\alpha(x))$ and $D(\beta\psi^{-1})(\psi f(x))$ are nonsingular matrices. But the rank of a matrix is unchanged if we pre- or postmultiply by a nonsingular matrix. Thus $D(\beta f\alpha^{-1})(\alpha(x))$ and $D(\psi f\varphi^{-1})(\varphi(x))$ have the same rank. □

Definition Let $f : M^m \to N^n$ be a differentiable function and let $x \in M$. The *rank* of f at x is the rank of $\psi f\varphi^{-1}$ at $\varphi(x)$, where (U,φ) and (V,ψ) are charts in the structures of M and N, and $x \in U$, $f(x) \in V$.

Lemma 4 assures us that the rank of f at x is well defined, i.e., independent of the choice of charts.

Definition The function $f : M^m \to N^n$ is a C^r *immersion* iff it is a C^r function having rank m at each point of M (thus $m \leq n$). If $f : M \to N$ is a C^r immersion which carries M homeomorphically onto $f(M)$ then f is a C^r *embedding*. If, in addition, f is a homeomorphism, then we say that f is a C^r *diffeomorphism*. In the last case $m = n$, and by the inverse function theorem, f^{-1} is also C^r.

Note that the definition of C^r embedding is stronger than the topological definition given in Chap. 2 (even if we require the topological embedding to be C^r) since we require further that the function have maximal rank. Use of the word "embedding" in the differential context will imply this stronger sense unless qualified by the adjective "topological."

If $f : M \to N$ is an embedding, then f(M) inherits a natural differential structure from M making f(M) into a differentiable manifold. Note that, by Theorem 3.6, a C^r function with compact domain is an embedding iff it is an injective immersion.

Just as homeomorphic spaces are usually considered the same in topology, diffeomorphic manifolds are usually considered the same in differential topology. Two problems spring to mind: Does every topological manifold support a differential structure? Although a manifold may support many different differential structures, are all

5. Differentiable Manifolds

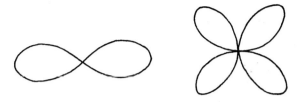

FIGURE 21

of the resulting manifolds diffeomorphic? The answer is yes to both questions for dimension at most 3, but no in general. For example, there is a topological 6-manifold which cannot be given any differential structure. On the other hand, S^7 has exactly 28 nondiffeomorphic structures.

Examples $f : \mathbb{R} \to \mathbb{R}$ given by $f(x) = x^3$ is a C^r homeomorphism, but, having rank 0 at 0, is not an immersion, hence not a C^r embedding or diffeomorphism.

The function $S^1 \to \mathbb{R}^2$ given by $e^{i\theta} \mapsto \sin 2\theta \, e^{i\theta}$ immerses S^1 in $\mathbb{R}^2$, the image being a four-petaled rose (see Fig. 21). Fig. 21 also depicts S^1 immersed in $\mathbb{R}^2$ so that the image is a figure eight. The Klein bottle immerses, but does not embed, in $\mathbb{R}^3$; the usual picture of a Klein bottle depicts one such immersion. It does imbed in $\mathbb{R}^4$. In fact, any compact m-manifold embeds in $\mathbb{R}^{2m}$, and for $m > 1$, immerses in $\mathbb{R}^{2m-1}$.

EXERCISES

1. Construct a natural C^∞ structure on the ellipsoid

$$\frac{x^2}{a^2} + \frac{y^2}{b^2} + \frac{z^2}{c^2} = 1$$

Is this manifold diffeomorphic to S^2? Explain.

2. Prove that the composition of two C^r functions is a C^r function and that the composition of two C^r diffeomorphism is a C^r diffeomorphism.

3. Let M be a manifold. Show that any C^{r+1} structure for M is a basis for a C^r structure. Construct, for some M and r, two distinct C^{r+1} structures which are bases for the same C^r structure.

4. Construct infinitely many different differential structures on a compact manifold.

5. Let $(M^m, \mathcal{D})$ be a differentiable manifold, let x and y be distinct points of M and let u,v be distinct points of $\mathbb{R}^m$. Prove that there exists $(U,\varphi) \in \mathcal{D}$ such that $x,y \in U$ and $\varphi(x) = u$, $\varphi(y) = v$. (Hint: Refer to Lemma 1.)

6. Let $(M^m, \mathcal{D})$ be a differentiable manifold. Prove that M is connected iff for all $x,y \in M$, there exists $(U,\varphi) \in \mathcal{D}$ with $x,y \in U$ and $\varphi(U) = \mathbb{R}^m$. (Hint: Refer to exercise 3.8.)

7. Let M^m be a topological manifold and suppose $f : \mathbb{R}^m \to M$ is a continuous surjection satisfying the following condition: For all $y \in M$ and all $x \in f^{-1}(y)$, there exists open $U_x \subset \mathbb{R}^m$ such that $x \in U_x$, $f \mid U_x$ is injective, and if $x,x' \in f^{-1}(y)$ then $f(U_x) = f(U_{x'})$ and

$$U_x \xrightarrow{f \mid U_x} f(U_x) = f(U_{x'}) \xrightarrow{(f \mid U_{x'})^{-1}} \mathbb{R}^m$$

is a C^r function. Construct a natural differential structure of class C^r on M using the function f. Under this differential structure, f should be a C^r immersion. [The existence of such a function f is considered in Ref. 4, where it is shown that, under mild hypotheses on M (including connectedness), such a function always exists, at least if we relax the requirement that $(f \mid V)^{-1} (f \mid U)$ be C^r.]

6
ORIENTABILITY

This chapter is an odd one out in the sense that the main concept introduced here is not really needed until Chap. 13. It is introduced here partly because it is a relatively familiar concept and partly because a later appearance would interrupt the flow of the other concepts.

The main concept of this chapter is an abstraction of the one- or two-sidedness of an object and also the right- or left-handedness of an object. For example the plane $\mathbb{R}^2$ is two-sided whereas the Möbius strip, which is a manifold if we omit the edge, is only one-sided. There are also compact manifold examples: S^2 is two-sided whereas the Klein bottle has only one side. Unfortunately this way of looking at these manifolds is unsuitable in the sense that it involves not only the manifold, but also an embedding (or at least an immersion) in euclidean space. The choice of a different embedding might cause problems. Thus when we look at S^1 embedded in $\mathbb{R}^2$ in the standard way we would declare S^1 to be two-sided, but the situation is rather changed if S^1 is embedded in $\mathbb{R}^3$. In fact it does not really make much sense to consider the one- or two-sidedness of a manifold unless it is embedded in euclidean space of one higher dimension. It is better for our definition to be an intrinsic one, i.e., refer only to the manifold itself and not to a particular way it is embedded in euclidean space (if indeed it can be!). Right- and left-handedness are much more intrinsic, and it is this which we are able to adapt to manifolds. We begin by settling the situation in euclidean space and find that it is then quite easy to transfer to manifolds.

The chapter is completed by a description of a family of manifolds, one in each dimension. As a source of examples these manifolds will be useful. They will also illustrate another way of constructing manifolds.

The simplest way to turn a right hand into a left hand is to look at it in a mirror. If the mirror is placed in $\mathbb{R}^3$ as the plane with equation $x = 0$, then looking into it determines the diffeomorphism of $\mathbb{R}^3$ given by $(x,y,z) \mapsto (-x,y,z)$. The jacobian determinant of this diffeomorphism is negative. A rotation in $\mathbb{R}^3$ will not turn a right hand into a left hand, and one can check that the (jacobian) determinant of a rotation is positive. A general diffeomorphism of $\mathbb{R}^2$ or $\mathbb{R}^3$ will convert a right-handed coordinate system about a point into a curvilinear system, a right-handed system if the jacobian determinant is positive and a left-handed system if the jacobian determinant is negative. Thus we arrive at the following definition.

Definition Let $f : U \to V$ be a diffeomorphism between open subsets of $\mathbb{R}^n$. Say that f is *orientation preserving (reversing)* at $x \in U$ iff the jacobian determinant of f at x is positive (negative), and say that f is *orientation preserving (reversing)* iff for all $x \in U$, f is orientation preserving (reversing) at x. Recall that the jacobian determinant of a diffeomorphism is never 0.

Examples The identity function and any translation are orientation preserving. Any reflection through a hyperplane of dimension n - 1 is orientation reversing, e.g., $f : \mathbb{R}^3 \to \mathbb{R}^3$ defined by $f(x,y,z) = (-y,-x,z)$ is reflection in the plane $x + y = 0$.

$$Df(x,y,z) = \begin{pmatrix} 0 & -1 & 0 \\ -1 & 0 & 0 \\ 0 & 0 & 1 \end{pmatrix}$$

6. Orientability

so $\Delta f(x,y,z) = -1 < 0$ at each point of $\mathbb{R}^3$.

LEMMA 1. *Let $f : U \to V$ be a diffeomorphism which is orientation preserving at some point of U. If U is connected, then f is orientation preserving.*

Proof. Define $\sigma : U \to \mathbb{R}$ by $\sigma(x) = \Delta f(x)/|\Delta f(x)|$. Then $\sigma(x)$ is, effectively, the sign of the jacobian determinant at x. Note that $\sigma(U) \subset \{-1,1\}$. Further, since f is orientation preserving at some point, $\sigma^{-1}(1) \neq \phi$. Moreover, U is connected and σ is continuous (since f is C^1). Thus $\sigma(U) = \{1\}$, so $\Delta f(x) > 0$ for all $x \in U$, i.e., f is orientation preserving. □

Definition A differentiable manifold $(M, \mathcal{D})$ is *orientable* iff there is a basis $\mathcal{B}$ for $\mathcal{D}$ such that

Orient For all $(U, \varphi), (V, \psi) \in \mathcal{B}$, the coordinate transformation $\psi \varphi^{-1}$ is orientation preserving.

An *orientation* of $(M, \mathcal{D})$ is a basis which is maximal with respect to Orient. An *oriented manifold* is a pair $(M, \mathcal{B})$ where $\mathcal{B}$ is an orientation for the manifold M.

If $f : M \to N$ is a diffeomorphism between two oriented manifolds, say $\mathcal{B}$ and $\mathcal{C}$ are the respective orientations, then we say that f is *orientation preserving (reversing)* provided $\psi f \varphi^{-1}$ is, for all $(U, \varphi) \in \mathcal{B}$ and all $(V, \psi) \in \mathcal{C}$.

Examples S^m and T^2 with their usual structures are orientable, but the Möbius strip is not. To verify the nonorientability of the Möbius strip, we can appeal to Theorem 3 below. To verify orientability of S^m and T^2, we take the bases for the usual structures of these manifolds constructed in Chap. 4. In fact the basis for T^2 constructed in Chap. 5 is a basis for an orientation, since, as determined there, on the annulus between the circles of radii π and 2π, $\psi \varphi^{-1}$ is the identity, which is orientation preserving, and on the annulus between the circles of radii 2π and 3π,

$$\psi\varphi^{-1}(x,y) = \frac{\sqrt{x^2+y^2} - 2\pi}{\sqrt{x^2+y^2}}(x,y)$$

$$= \left(x - \frac{2\pi x}{\sqrt{x^2+y^2}},\ y - \frac{2\pi y}{\sqrt{x^2+y^2}}\right)$$

so

$$D\psi\varphi^{-1}(x,y) = \begin{pmatrix} 1 - 2\pi \dfrac{\sqrt{x^2+y^2} - x^2/\sqrt{x^2+y^2}}{x^2+y^2} & \dfrac{2\pi xy}{(x^2+y^2)^{3/2}} \\ \\ \dfrac{2\pi xy}{(x^2+y^2)^{3/2}} & 1 - 2\pi \dfrac{\sqrt{x^2+y^2} - y^2/\sqrt{x^2+y^2}}{x^2+y^2} \end{pmatrix}$$

$$= \begin{pmatrix} 1 - \dfrac{2\pi y^2}{(x^2+y^2)^{3/2}} & \dfrac{2\pi xy}{(x^2+y^2)^{3/2}} \\ \\ \dfrac{2\pi xy}{(x^2+y^2)^{3/2}} & 1 - \dfrac{2\pi x^2}{(x^2+y^2)^{3/2}} \end{pmatrix}$$

From this it follows that

$$\Delta\psi\varphi^{-1}(x,y) = 1 - \frac{2\pi}{\sqrt{x^2+y^2}} > 0 \qquad \text{if } \sqrt{x^2+y^2} > 2\pi$$

Thus $\psi\varphi^{-1}$ is orientation preserving, so T^2 is orientable.

To verify orientability of S^m, we need to modify the basis constructed in Chap. 5, since the coordinate transformation determined by that basis is orientation reversing (but see Theorem 3 below for a way of avoiding this minor problem).

Let

$$U = S^m - \{(0,\ldots,0,1)\} \qquad V = S^m - \{(0,\ldots,0,-1)\}$$

and define $\varphi : U \to \mathbb{R}^m$, $\psi : V \to \mathbb{R}^m$ by

6. Orientability

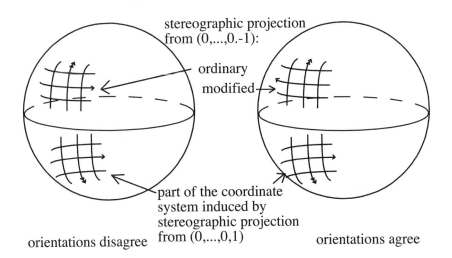

FIGURE 22

$$\varphi(x_0, \ldots, x_m) = \left(\frac{x_0}{1 - x_m}, \ldots, \frac{x_{m-1}}{1 - x_m} \right)$$

$$\psi(x_0, \ldots, x_m) = \left(-\frac{x_0}{1 + x_m}, \frac{x_1}{1 + x_m}, \ldots, \frac{x_{m-1}}{1 + x_m} \right)$$

These modified charts are illustrated in Fig. 22. Then (U,φ) is the stereographic projection of Chap. 3, and (V,ψ) is a slightly altered version of this projection. Modifying the calculation in Chap. 5 suitably, we obtain, in the present case,

$$\psi\varphi^{-1}(y_1, \ldots, y_m) = \left(-\frac{y_1}{\sum_{i=1}^m y_i^2}, \frac{y_2}{\sum_{i=1}^m y_i^2}, \ldots, \frac{y_m}{\sum_{i=1}^m y_i^2} \right)$$

so that

$$D\psi\varphi^{-1}(y_i) = \frac{1}{(\Sigma y_i^2)^2} \begin{pmatrix} -\Sigma y_i^2 + 2y_1^2 & 2y_1 y_2 & \cdots & 2y_1 y_m \\ -2y_1 y_2 & \Sigma y_i^2 - 2y_2^2 & \cdots & -2y_2 y_m \\ -2y_1 y_3 & -2y_2 y_3 & \cdots & -2y_3 y_m \\ \vdots & & & \vdots \\ -2y_1 y_m & -2y_2 y_m & \cdots & \Sigma y_i^2 - 2y_m^2 \end{pmatrix}$$

Letting $(y_1,\ldots,y_m) = (\pm 1, 0,\ldots,0) \in \varphi(U \cap V) = \mathbb{R}^m - \{0\}$, we get

$$D\psi\varphi^{-1}(\pm 1, 0,\ldots,0) = \frac{1}{1} \begin{pmatrix} -1+2 & 0 & \cdots & 0 \\ 0 & 1 & \cdots & 0 \\ 0 & 0 & \cdots & 0 \\ \vdots & \vdots & & \vdots \\ 0 & 0 & \cdots & 1 \end{pmatrix}$$

so that $\Delta\psi\varphi^{-1}(\pm 1, 0,\ldots,0) = 1 > 0$, i.e., $\psi\varphi^{-1}$ is orientation preserving at $(\pm 1, 0,\ldots,0)$. If $m > 1$ then $\varphi(U \cap V)$ is connected, so by Lemma 1, $\psi\varphi^{-1}$ is orientation preserving. If $m = 1$, the domain of $\psi\varphi^{-1}$ is $(-\infty, 0) \cup (0, \infty)$. Since $(-\infty, 0)$ and $(0, \infty)$ are each connected, and $\psi\varphi^{-1}$ is orientation preserving at a point of each, again by Lemma 1, $\psi\varphi^{-1}$ is orientation preserving. Thus in all cases, S^m is orientable.

We now work toward a useful criterion for orientability.

LEMMA 2. Let $\rho : \mathbb{R}^m \longrightarrow \mathbb{R}^m$ be the reflection given by $\rho(x_1,\ldots,x_m) = (-x_1, x_2,\ldots,x_m)$. Thus ρ is reflection in the $x_2 \ldots x_m$ hyperplane. Suppose B is an orientation for the orientable manifold $(M^m, \mathcal{D})$. Let $(U,\varphi) \in \mathcal{D}$ be such that U is connected. Then either $(U,\varphi) \in B$ or $(U, \rho\varphi) \in B$ (but not both!).

6. Orientability

Proof: Let

$$U_0 = \{x \in U \mid \text{ for all } (V,\psi) \in B, \text{ either } x \notin V \text{ or } \Delta(\varphi\psi^{-1})(\psi(x)) > 0\}$$

In effect, U_0 is that part of U on which φ is orientably compatible with the orientation B. For $(V,\psi) \in B$, if $x \notin V$ then there is no incompatibility, but if $x \in V$ we require the coordinate transformation to be orientation preserving.

It is claimed that U_0 is open in M. For suppose $x \in U_0$. Since $x \in U$, U is open, and B is maximal, by exercise 3.9, there exists $(W,X) \in B$ such that W is connected and $x \in W \subset U$. It suffices to show that $W \subset U_0$, for then we will have found, for each $x \in U_0$, an open neighborhood of x contained in U_0, so U_0 is a neighborhood of each of its points.

Let $y \in W$ and $(V,\psi) \in B$ be such that $y \in V$. We must show that $\Delta(\varphi\psi^{-1})(\psi(y)) > 0$ to be able to deduce that $y \in U_0$. In $V \cap W$, $\varphi\psi^{-1} = \varphi X^{-1} X \psi^{-1}$, so that

$$\Delta(\varphi\psi^{-1})(\psi(y)) = \Delta(\varphi X^{-1})(X(y)) \times \Delta(X\psi^{-1})(\psi(y))$$

Since $(V,\psi), (W,X) \in B$, we must have $\Delta(X\psi^{-1})(\psi(y)) > 0$. Since $x \in U_0$ and $(W,X) \in B$, we have $\Delta(\varphi X^{-1})(X(x)) > 0$; thus φX^{-1} is orientation preserving at $X(x)$. Since $X(W)$ is connected, by Lemma 1, φX^{-1} is orientation preserving; in particular, at $X(y)$, i.e., $\Delta(\varphi X^{-1})(X(y)) > 0$. Since both $\Delta(\varphi X^{-1})(X(y))$ and $\Delta(X\psi^{-1})(\psi(y))$ are positive, so must $\Delta(\varphi\psi^{-1})(\psi(y))$ be.

Similarly the set

$$U_1 = \{x \in U \mid \text{ for all } (V,\psi) \in B, \text{ either } x \notin V \text{ or } \Delta(\varphi\psi^{-1})(\psi(x)) < 0\}$$

is open in M.

Clearly $U_0 \cap U_1 = \phi$. Moreover, $U_0 \cup U_1 = U$, for if $x \in U$ but $x \notin U_0$, then there exists $(W,X) \in B$ with $x \in W$ but $\Delta(\varphi\psi^{-1})(X(x)) \not> 0$. Since φX^{-1} is a diffeomorphism, we cannot have $\Delta(\varphi X^{-1})(X(x)) = 0$, so $\Delta(\varphi X^{-1})(X(x)) < 0$. Now if $(V,\psi) \in B$ and $x \in V$, then $\Delta(X\psi^{-1})(\psi(x)) > 0$, so $\Delta(\varphi\psi^{-1})(\psi(x)) < 0$, being the product of the negative number $\Delta(\varphi\chi^{-1})(\chi(x))$ and the positive number $\Delta(\chi\psi^{-1})(\psi(x))$. Hence $x \in U_1$.

Since U is connected, by Corollary 2.5 either $U_0 = \phi$ or $U_1 = \phi$. If $U_0 = \phi$, then $U_1 = U$, hence

$$\Delta(\varphi\psi^{-1})(\psi(x)) < 0 \qquad \text{for all } (V,\psi) \in B, \text{ for all } x \in U \cap V$$

so

$$\Delta(\rho\varphi\psi^{-1})(\psi(x)) = \Delta(\rho)(\varphi(x)) \times \Delta(\varphi\psi^{-1})(\psi(x)) > 0$$

Thus, since ρ is a diffeomorphism, so that $(U,\rho\varphi) \in \mathcal{D}$, by maximality of B we must have $(U,\rho\varphi) \in B$. If, on the other hand, $U_1 = \phi$, then $U_0 = U$ and by maximality of B, $(U,\varphi) \in B$. □

THEOREM 3. Let $(M,\mathcal{D})$ be a differentiable manifold. Then $(M,\mathcal{D})$ is orientable if and only if for all $(U,\varphi),(V,\psi) \in \mathcal{D}$ for which U and V are connected, $\Delta(\varphi\psi^{-1})$ has constant sign throughout $\psi(U \cap V)$.

Proof: Suppose that $(M,\mathcal{D})$ is orientable, say B is an orientation of M. Let $(U,\varphi),(V,\psi) \in \mathcal{D}$ with U and V connected. By Lemma 2 either (U,φ) or $(U,\rho\varphi)$ and either (V,ψ) or $(V,\rho\psi)$ belongs to B. Hence one of $\varphi\psi^{-1}$, $\rho\varphi\psi^{-1}$, $\varphi\psi^{-1}\rho^{-1}$, or $\rho\varphi\psi^{-1}\rho^{-1}$ has positive jacobian determinant throughout $\psi(U \cap V)$ or $\rho\psi(U \cap V)$. Since $\rho^{-1} = \rho$ and ρ is orientation reversing, we deduce, respectively, that $\Delta(\varphi\psi^{-1})$ is positive, negative, negative, or positive throughout $\psi(U \cap V)$. In any case, the sign is constant.

Conversely, suppose that the constant sign condition holds. We will construct a basis for an orientation of $(M,\mathcal{D})$, i.e., a basis for $\mathcal{D}$ satisfying Orient. The orientations of different components of M need bear no relationship with one another, so we assume that M is connected. If it is not then carry out the following process on each component separately.

Pick $(U,\varphi) \in \mathcal{D}$ for which U is connected and nonempty, say $x \in U$. In effect, (U,φ) determines the orientation of $(M,\mathcal{D})$. Let

$B = \{(V,\psi) \in \mathcal{D} \mid V \text{ is connected, } x \in V \text{ and } \psi\varphi^{-1} \text{ is orientation preserving on } \varphi(U \cap V)\}$

6. Orientability

It is claimed that B is a basis for an orientation of $(M,\mathcal{D})$.

By Exercise 5.6, for all $y \in M$, there exists $(V,\psi) \in \mathcal{D}$ such that $y \in V$, V is connected, and $x \in V$. By hypothesis, $\Delta(\psi\varphi^{-1})$ has constant sign throughout $\varphi(U \cap V)$. If this sign is positive then $(V,\psi) \in B$. If the sign is negative, then $(V,\rho\psi) \in B$. In either case we have found, for all $y \in M$, a chart $(V,\psi) \in B$ such that $y \in V$. Thus B is a basis for $\mathcal{D}$.

Suppose $(V,\psi),(W,\chi) \in B$. We show that B satisfies Orient by verifying that $\chi\psi^{-1} : \psi(V \cap W) \to \chi(V \cap W)$ is orientation preserving. In fact, at $\psi(x)$, $\chi\psi^{-1} = \chi\varphi^{-1}\varphi\psi^{-1}$, so, being a composition of orientation-preserving diffeomorphisms, $\chi\psi^{-1}$ is orientation preserving at $\psi(x)$; hence $\chi\psi^{-1}$ is orientation preserving, by hypothesis. Thus $(M,\mathcal{D})$ is orientable. □

There is a major difference between our two criteria for orientability, viz., the definition begins with the quantifier $\exists$ whereas the criterion of Theorem 3 begins with $\forall$. To verify that a manifold is orientable using the definition, we must exhibit a basis satisfying a certain condition--not usually a very daunting task, particularly if the manifold is compact so that only finitely many charts are needed for a basis. This is how we verified the orientability of S^m and T^2. On the other hand, to verify that a manifold is orientable using Theorem 3, we must check a certain condition for all possible pairs from an infinite family. There may be clever ways to reduce the size of the family, but in general this is not a recommended way of verifying orientability.

If we negate the two criteria for orientability then we obtain criteria for nonorientability. From the definition we obtain a criterion beginning with the quantifier $\forall$ whereas Theorem 3 yields a criterion beginning with $\exists$. Thus the latter criterion will probably be the more useful. We will use this criterion to verify the nonorientability of P^2 below.

We complete this chapter by describing an important family of manifolds, the (real) *projective spaces*. Let

$$P^n = \{\{x,-x\} \mid x \in S^n\}$$

and define $\pi : S^n \longrightarrow P^n$ by $\pi(x) = \{x,-x\}$. Then the set P^n consists of all *unordered* pairs of antipodal points of S^n, and π takes a point of S^n and pairs it with its antipodal point. Note that $\pi(x) = \pi(-x)$, since $\{x,-x\} = \{-x,x\}$. Technically, we say that P^n is the identification space obtained by identifying antipodal points of S^n. The set P^n is topologized by declaring open those sets of the form $\pi(U)$ where U is an open subset of S^n. This makes π into a continuous function. In fact, if $V \subset P^n$, then V is open iff $\pi^{-1}(V)$ is open. Thus, since π is injective on any open hemisphere of S^n, π is actually an embedding of an open hemisphere of S^n in P^n. Because of this, P^n is a differentiable n-manifold. More precisely, let $\mathcal{B}$ be a basis for the usual differential structure on S^n such that for all $(U,\varphi) \in \mathcal{B}$, U lies entirely inside some open hemisphere of S^n. Then

$$\{(\pi(U), \varphi\pi^{-1}) \mid (U,\varphi) \in \mathcal{B}\}$$

is a basis for a differential structure on P^n. (Why is P^n Hausdorff?) Note that P^n is compact.

Recall that S^{n-1} is treated as a subset of S^n by adding on the last coordinate of 0. Since π respects this inclusion, i.e., if $x,y \in S^n$ are such that $\pi(x) = \pi(y)$ then $x \in S^{n-1}$ iff $y \in S^{n-1}$, we obtain an inclusion of P^{n-1} in P^n, and a commutative diagram:

$$S^0 \hookrightarrow S^1 \hookrightarrow S^2 \hookrightarrow S^3 \hookrightarrow \cdots$$
$$\downarrow \pi \quad \downarrow \pi \quad \downarrow \pi \quad \downarrow \pi \quad \downarrow \cdots$$
$$P^0 \hookrightarrow P^1 \hookrightarrow P^2 \hookrightarrow P^3 \hookrightarrow \cdots$$

Now $S^0 = \{1,-1\}$, so $P^0 = \{\{1,-1\}, \{-1,1\}\} = \{\{1,-1\}\}$, i.e., P^0 consists of a single point (with the only possible topology!).

6. Orientability

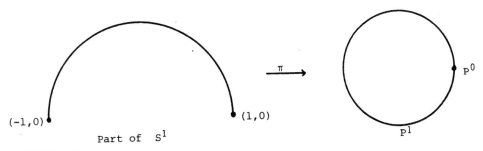

FIGURE 23

If we start moving around S^1 from $(1,0)$ in the counterclockwise direction toward $(-1,0)$, we obtain new points of P^1 until we reach $(-1,0)$: $\pi(1,0) = \pi(-1,0)$. Applying π to this upper semicircle of S^1 gives us the whole of P^1 together with its topological structure, an interval with its endpoints identified. This gives us S^1 again (up to homeomorphism), as shown in Fig. 23.

The lower semicircle S^1 gives us nothing new; π merely takes us around P^1 again, i.e., π wraps S^1 around P^1 twice (see Fig. 24).

The above description of P^1 suggests another way of defining P^n. We need only take the upper closed hemisphere and identify antipodal points on the boundary. Since the upper closed hemisphere is homeomorphic to B^n, we can obtain P^n from B^n by identifying antipodal points on the boundary S^{n-1}. Since P^{n-1} is obtained from S^{n-1} by identifying antipodal points, we see that P^n is obtained

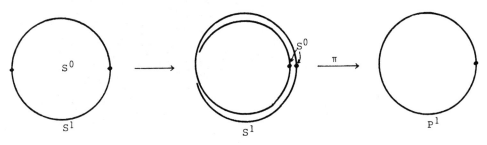

FIGURE 24

from P^{n-1} by attaching an open disk. To obtain the topological structure of P^n, we prefer to keep all of B^n rather than throw away "superfluous" points of S^{n-1}; cf. Fig. 23 in which the topological structure of P^1 above P^0 comes from the structure of S^1 above $(1,0)$, whereas the structure of P^1 below P^0 comes from the structure of S^1 above $(-1,0)$. Had we removed $(-1,0)$ as well, then we would not have known to bend that part of S^1 just above $(-1,0)$ around to P^0. We would not have known that $\pi(1,0)$ is near $\pi(A)$, where

$$A = \{(x,y) \in S^1 \mid x < 0, y < 0\}$$

Using the above view of P^n, we can obtain three pictures of P^2, the first rather symbolic. Fig. 25 shows B^2; to obtain P^2 we must identify antipodal points on the boundary, so the two points labeled a are the same, as are the two points labeled b and the two points labeled c. (The ball B^2 could also be thought of as providing a model of S^2 provided we identify all points on the boundary, a kind of reversal of stereographic projection obtained by tightening a drawcord running all the way around the boundary.) The two lines joining a and c when put together represent a circle embedded in P^2. This circle bounds two regions in P^2 each of which may be constructed separately; then the two regions may be joined together along the circle. What do these two regions look like? Figure 25 suggests

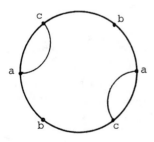

FIGURE 25

6. *Orientability* 77

that there are three regions: two similar lens-shaped regions and a curvilinear quadrilateral. However, since antipodal points on the boundary are the same, the two lens-shaped regions are really parts of one region, a disk obtained by identifying the relevant parts of the boundary. The identification necessary on the curvilinear quadrilateral is easy and it results in a Möbius strip. Thus P^2 is obtained by gluing together a Möbius strip and a disk along their edges (each edge is homeomorphic to S^1; cf. limerick II of Chap. 1).

The reader might like to construct a model of P^2 by gluing together a Möbius strip and a disk, as described in the previous paragraph. Fig. 26 suggests a way of doing this. On the left is a Möbius strip. Its bounding circle is shaded in such a way that the portions nearest the reader are thicker. By rotating this bounding circle around a horizontal line across the page so that the nearer parts move sufficiently up the page to eliminate the center loop, we obtain the circle in the right picture; this picture shows this circle bounding a disk. Combining the two gives us our model of P^2. In order to combine the two, we must introduce a line of self-intersection, say L and L' as shown. This line of self-intersection is definitely not an intrinsic part of P^2, but rather a shortcoming of our model, a shortcoming which we cannot completely eliminate as P^2 does not embed in $\mathbb{R}^3$. As long as we are aware of

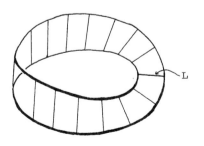 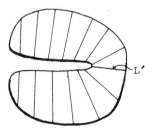

FIGURE 26

and respect this shortcoming, we may use the model freely. If we are moving around the center of the Möbius strip and come to the line of self-intersection, we should continue as if the other sheet were not there. We cannot move from the Möbius strip to the disk except by crossing the common bounding circle.

The third picture of P^2 is really the same as the second one. To obtain this picture, one distorts the disk, as shown in Fig. 27, so that antipodal points on the bounding circle become the same. In so doing, we again introduce a line of self-intersection. The resulting map of P^2 in $\mathbb{R}^3$ is an immersion except at the endpoints of the segment of self-intersection.

If we remove a nice disk from the bottom of the surface in Fig. 27, we are left with what is called a *cross-cap*, an important building block used in the classification of nonorientable surfaces.

P^2 does immerse in $\mathbb{R}^3$. Boy's surface provides an example of such an immersion. A method of constructing Boy's surface together with pictures of a wire model of it may be found in Ref. 9.

We now use Theorem 3 to show that P^2 is not orientable. Choose connected sets U and V as in Fig. 28. Then $U \cap V$ consists of two parts. However we choose $\varphi : U \to \mathbb{R}^2$ and $\psi : V \to \mathbb{R}^2$ in the structure of P^2, the diffeomorphism $\varphi \psi^{-1}$ will be orientation preserving on one part and reversing on the other, so by Theorem 3, P^2 is not orientable.

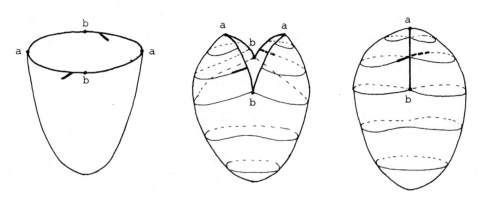

FIGURE 27

6. Orientability

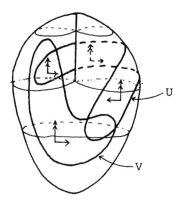

FIGURE 28

In fact, for $n > 0$, P^n is orientable iff n is odd. The non-orientability of P^{2n} follows in much the same way as for P^2: transfer the charts of Fig. 28 to Fig. 25 to get Fig. 29. If we coordinatize V as shown (the single arrow shows the x_1 axis, the double arrow the x_2 axis) and we coordinatize the left part of U to orientably agree with that of V, then on moving from the left part of U to the right part, the x_1 axis still points to the right, but the x_2 axis flips, to point downwards, because of the way the two

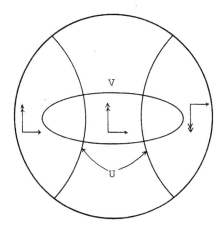

FIGURE 29

parts of U are glued together to get P^2. Extending this idea to P^n, we see that the x_2, x_3, ..., x_n axes all flip. When n is even, an odd number of axes flip to give us a reversal of orientation, so the same two charts show that P^n is not orientable for n even. When n is odd, an even number of axes flip and the orientation is not reversed. However, this does not show that in this case P^n is orientable, for Theorem 3 requires much more than a constant sign for the jacobian determinant of the coordinate transformation between two particular charts.

EXERCISES

1. Prove that the function $f : \mathbb{C} \to \mathbb{C}$ given by $f(z) = z^2$ determines an orientation-preserving diffeomorphism of the open upper half plane to the plane minus the nonnegative real axis. The orientations of domain and range should agree. If they do not then the diffeomorphism will be orientation reversing.
2. In Exercise 4.2 there was constructed a diffeomorphism between $\mathbb{R}^3$ and $\mathbb{R}^3 - \{(x,0,z) \mid x \geq 0\}$. Determine whether this diffeomorphism is orientation preserving.
3. Prove that a connected orientable manifold has exactly two orientations. What if "connected" is removed?
4. Let $(M^m, \mathcal{D})$ be a differentiable manifold. Let N be an open subset of M. Show that $E = \{(U \cap N, \varphi \mid U \cap N) \mid (U,\varphi) \in \mathcal{D}\}$ is a differential structure on N. Prove that if (N,E) is not orientable then $(M,\mathcal{D})$ is not orientable. Hence provide an alternative proof that P^2 is not orientable.
5. Use Exercise 5.7 to provide another way of constructing a differential structure on P^n. Is this the same differential structure as was constructed in the text?
6. Why is the Klein bottle not orientable?

7
SUBMANIFOLDS AND AN EMBEDDING THEOREM

This chapter has several purposes. The two overt purposes are to carry over to manifolds the nice way that $\mathbb{R}^m$ sits inside $\mathbb{R}^n$ ($m < n$) and to show that any compact manifold can be embedded in $\mathbb{R}^n$ for sufficiently large n. In the introduction to Chap. 4, we noted that surfaces are to be classified by looking at smooth real-valued functions defined thereon and studying how preimages of points vary. We will find in Chap. 9 that point preimages will usually be submanifolds, i.e., will sit inside the surface much as $\mathbb{R}^m$ sits inside $\mathbb{R}^n$.

This chapter also has a number of covert purposes, particularly in the proof of Theorem 2. For example, we will see how DS2 of Chap. 5 can be used to serve us. We will see how the finite character of compactness can be exploited. Perhaps most important, for this technique will be used again in more complicated situations, we will see how Lemma 4.1 can be used to extend the domain of a function to an entire manifold.

Suppose that M^m and N^n are manifolds and that $M \subset N$. Let $x \in M$. Then there are many charts in the structure of N about x. Most, if not all, will ignore the presence of M. If it happens that at each point $x \in M$ there is a chart in the structure of N with respect to which M lies in the same way as $\mathbb{R}^m$ does in $\mathbb{R}^n$, then this is something special. Fig. 30 illustrates the possibilities.

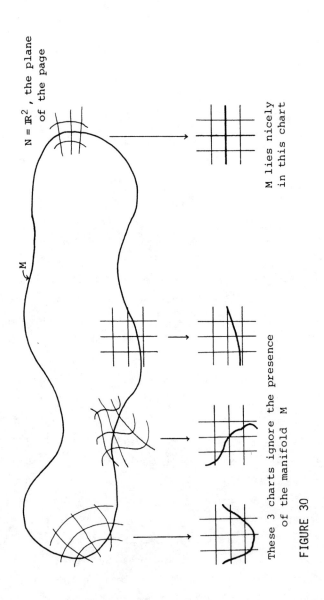

FIGURE 30

7. Submanifolds and an Embedding Theorem

Definition Let (M^m, D) and (N^n, E) be C^r manifolds for some r, with $M \subset N$. Say that M is a *submanifold* of N iff for all $x \in M$ there exists $(U, \varphi) \in E$ such that U is a neighborhood of x in N, $\varphi^{-1}(\mathbb{R}^m) = U \cap M$, and $(U \cap M, \varphi|U \cap M) \in D$.

The condition $\varphi^{-1}(\mathbb{R}^m) = U \cap M$ asserts two things. Under φ, the entire part of M lying in U is carried into $\mathbb{R}^m$; the only part of U which is carried to $\mathbb{R}^m$ by φ is that part lying in M. This means that if we restrict the chart (U, φ) to M then we get a chart $(U \cap M, \varphi|U \cap M)$ on M. The last condition demands that this chart lie in the differential structure of M. This condition is harder to represent pictorially, but it says, for example, that when we are considering the differentiability of a function with range M, it does not matter whether we use D or E.

Examples S^m is a submanifold of $\mathbb{R}^n$ $(n > m)$. To prove this, it is enough to consider the case where $n = m + 1$. Suppose given $\bar{x} = (\bar{x}_1, \ldots, \bar{x}_{m+1}) \in S^m$. At least one of the coordinates $\bar{x}_i$ is nonzero, say $\bar{x}_{m+1} > 0$. As is illustrated by Fig. 31, we obtain the required embedding φ by pushing the hemisphere consisting of all points of S^m whose last coordinate is positive to $\mathbb{R}^m$ (recall that $\mathbb{R}^m$ is the set of points of $\mathbb{R}^{m+1}$ with last coordinate 0). Let

$$U = \{(x_1, \ldots, x_{m+1}) \in \mathbb{R}^{m+1} \mid \sum_{i=1}^{m} x_i^2 < 1 \text{ and } x_{m+1} > 0\}$$

Then U is that part of $\mathbb{R}^{m+1}$ obtained by translating $\text{Int } B^m \subset \mathbb{R}^m$ a positive amount in the $(m + 1)$st direction. Note that $\text{Int } B^m$ is precisely that portion of $\mathbb{R}^m$ bounded by $S^m \cap \mathbb{R}^m = S^{m-1}$. The set U is an open subset of $\mathbb{R}^{m+1}$ and contains all points of S^m with positive last coordinate; in particular, $\bar{x} \in U$. Now define $\varphi : U \to \mathbb{R}^{m+1}$ so that φ moves each point of U in the $(m + 1)$st direction, moving points of $S^m \cap U$ onto $\mathbb{R}^m$; thus φ translates each line parallel to the x_{m+1} axis by an amount equal to the x_{m+1} coordinate of the point on that line lying on S^m. Precisely, let

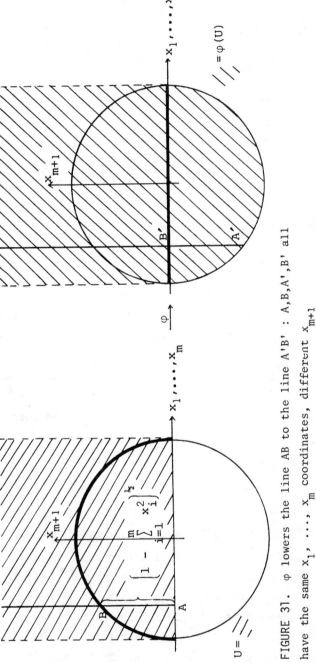

FIGURE 31. φ lowers the line AB to the line A'B' : A,B,A',B' all have the same $x_1, \ldots, x_m$ coordinates, different x_{m+1} coordinates.

7. Submanifolds and an Embedding Theorem

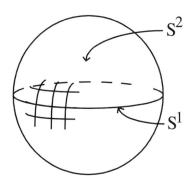

FIGURE 32

$$\varphi(x_1,\ldots,x_{m+1}) = \left[x_1,\ldots,x_m,x_{m+1} - \left(1 - \sum_{i=1}^m x_i^2\right)^{\frac{1}{2}}\right]$$

Clearly φ is a C^∞ embedding since $\sum_{i=1}^m x_i^2 < 1$, $\varphi^{-1}(\mathbb{R}^m) = U \cap S^m$, and one can check that $(U \cap S^m, \varphi|U \cap S^m)$ is in the structure of S^m (using the basis considered in Chap. 5). If $x_{m+1} < 0$, the above chart can be modified to give us a chart required by the definition: replace U by its reflection in the $x_1 \ldots x_m$ hyperplane and slide lines upward instead of downward.

One can show that S^m is a submanifold of S^n ($m \le n$). Lines of latitude and longitude near the equator provide charts of the required form when $m = 1$, $n = 2$ (see Fig. 32). T^2 is a submanifold of $\mathbb{R}^3$.

Clearly if M is a submanifold of N then the inclusion function $M \hookrightarrow N$ is an embedding. A suitably framed converse also holds.

THEOREM 1. Let $e : M^m \longrightarrow N^n$ be an embedding. If $e(M)$ has the differential structure inherited from M by e, $e(M)$ is a submanifold of N.

Proof: Let $\mathcal{D}$ and $\mathcal{E}$ be the differential structures of M and N, respectively. Let

$$e(\mathcal{D}) = \{(e(U), \varphi e^{-1}) \mid (U,\varphi) \in \mathcal{D}\}$$

The reader should check that $e(\mathcal{D})$ is a differential structure on $e(M)$; this is the structure referred to in the statement of the theorem.

Let $p \in M$. By differentiability of e, there exists $(V,\psi) \in \mathcal{D}$, $(W,\chi) \in E$ such that $p \in V$, $e(V) \subset W$, and $\chi e \psi^{-1}$ is differentiable. By Lemma 5.1, we may assume that $\psi(p) = 0 = \chi e(p)$. The chart (W,χ) will be modified to give the required chart about $e(p)$.

Since e is an embedding, it has rank m, so $\chi e \psi^{-1}$ also has rank m. Applying Corollary 4.3 to $\chi e \psi^{-1}$, we obtain a diffeomorphism g of a neighborhood of 0 in $\mathbb{R}^n$ onto another such neighborhood such that

(*) $\quad g\chi e\psi^{-1}(x_1,\ldots,x_m) = (x_1,\ldots,x_m,0,\ldots,0)$

Since $\chi(W)$ is a neighborhood of 0, we may assume that $\chi(W)$ contains the domain of g. We may also assume that V is small enough so that equation (*) is valid throughout $\psi(V)$. Choose a positive number r small enough so that $rB^m \subset g\chi e(V)$, $rB^n \subset$ image of g, and $rB^n \cap g\chi e(M) = rB^m$, and let $U = \chi^{-1}g^{-1}(\text{Int } rB^n)$ and $\varphi = g\chi$. It may be that W sprawls to meet $e(M)$ far from $e(p)$ and χ might carry this remote part of $e(M)$ into the domain of g. However, g might not straighten this part of $e(M)$ as in equation (*). By choosing r so small that $rB^n \cap g\chi e(M) = rB^m$, we effectively cut such pieces of $e(M)$ out of U.

Since $\chi e(p) = 0$, so that $g\chi e(p) = 0$, we have $e(p) \in U$. Further U is the inverse image of an open set so, by Theorem 2.4, U is a neighborhood of $e(p)$. Moreover,

$$\varphi^{-1}(\mathbb{R}^m) = \chi^{-1}g^{-1}(\text{Int } rB^m) \qquad \text{by definition of } \varphi$$
$$= U \cap e(M) \qquad \text{by choice of } r$$

Finally, $\varphi \mid U \cap e(M) = \psi e^{-1} \mid U \cap e(M)$, so that $(U \cap e(M), \varphi \mid U \cap e(M)) \in e(\mathcal{D})$. $\square$

7. Submanifolds and an Embedding Theorem

Remark The proof above can be modified easily to obtain, for an immersion $f : M^m \to N^n$, charts (V,ψ) about any point $p \in M$ and (U,φ) about $f(p)$ so that for all $(x_1,\ldots,x_m) \in \psi(V)$,

$$\varphi f \psi^{-1}(x_1,\ldots,x_m) = (x_1,\ldots,x_m,0,\ldots,0) .$$

THEOREM 2. Let M be a compact differentiable manifold. Then M embeds in $\mathbb{R}^n$ for sufficiently large n.

Proof: Suppose $e : M \to \mathbb{R}^n$ is a function for some n. We may split e into its component functions $e_1,\ldots,e_n : M \to \mathbb{R}$. As noted in Chap. 5, an embedding is an injective immersion. Thus in particular, for e to be an embedding, it must send distinct points of M to distinct points of $\mathbb{R}^n$, so at least one of the component functions $e_1,\ldots,e_n$ must take on different values at a given pair of distinct points of M. The idea of the proof, then, is to find n differentiable functions $e_1,\ldots,e_n : M \to \mathbb{R}$ such that for all $y \in M$ with $x \neq y$, there exists i such that $e_i(x) \neq e_i(y)$. If (U,φ) is a chart, then the components of φ, say $\varphi_1,\ldots,\varphi_m : U \to \mathbb{R}$, satisfy this requirement within U; the main problem is that their domains are only U, not the whole of M. We must somehow extend them. The technique used below to extend these functions will be used again in different contexts in later chapters.

For each $x \in M$, choose a chart (U_x,φ_x) about x with $\varphi_x(x) = 0$ and $B^m \subset \varphi_x(U_x)$. Lemma 5.1 allows us to choose φ_x so that $\varphi_x(x) = 0$. Since $\varphi_x(U_x)$ is a neighborhood of 0, it must contain some ball centered at 0. Using DS2 to replace φ_x by the composition of φ_x with a diffeomorphism of $\mathbb{R}^m$ which expands this ball onto B^m achieves the required situation. Initially our choice of V_x and W_x following will appear strange, but the wisdom of these choices will manifest itself when we come to check the injectivity and immersivity of the constructed function e. Let $V_x = \varphi_x^{-1}(\text{Int } B^m)$ and $W_x = \varphi_x^{-1}(\text{Int } \tfrac{1}{2} B^m)$. Then $\{W_x \mid x \in M\}$ is an open cover of M. By compactness, there is a finite subcover, say $\{W_1,\ldots,W_p\}$; index the corresponding φ, U, V accordingly.

Let $h : \mathbb{R}^m \to \mathbb{R}$ be the function of Lemma 4.1, and for $i = 1, \ldots, p$, defined differentiable functions $h_i : M \to \mathbb{R}$ by

$$h_i(x) = \begin{cases} h\varphi_i(x) & \text{if } x \in U_i \\ 0 & \text{if } x \in M - \text{Cl } V_i \end{cases}$$

The function h_i is well defined, for if $x \in U_i \cap (M - \text{Cl } V_i)$, then $\varphi_i(x) \in \mathbb{R}^m - B^m$, so $h\varphi_i(x) = 0$. The function h_i is differentiable, for if $x \in M$, then either $x \in U_i$, in which case h_i is the composition of two differentiable functions in the neighborhood U_i of x, or $x \notin U_i$, in which case h_i is the constant (hence differentiable) function 0 in the neighborhood $M - \text{Cl } V_i$ of x. Since differentiability is a local phenomenon, these two facts combine to give the differentiability of h_i, the point being that the two pieces of M on which h_i is constructed are open sets. One might think of h_i as being a height function, assigning to each point of W_i the height 1, and each point outside V_i the height 0; thus W_i is a plateau and $M - V_i$ is the sea (see Fig. 33).

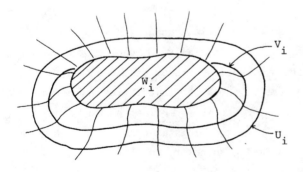

FIGURE 33

7. Submanifolds and an Embedding Theorem

Now "extend" φ_i over the manifold M using h_i to get $\hat{\varphi}_i : M \to \mathbb{R}^m$:

$$\hat{\varphi}_i(x) = \begin{cases} h_i(x)\varphi_i(x) & \text{if } x \in U_i \\ 0 & \text{if } x \in M - \text{Cl } V_i \end{cases}$$

Again $\hat{\varphi}_i$ is differentiable, being constant or a product of two differentiable functions. Further, since $h_i(\text{Cl } W_i) = 1$, we have $\hat{\varphi}_i \mid \text{Cl } W_i = \varphi_i \mid \text{Cl } W_i$. Thus $\hat{\varphi}_i$ extends only $\varphi_i \mid \text{Cl } W_i$; to obtain the extension, we have lost a bit of ground.

Define the required embedding

$$e : M \to (\mathbb{R}^m \times \mathbb{R})^p = \mathbb{R}^{mp+p}$$

by setting

$$e(x) = \big((\hat{\varphi}_1(x), h_1(x)), \ldots, (\hat{\varphi}_p(x), h_p(x))\big)$$

e is injective: Suppose $x, y \in M$ and $x \neq y$. Since $\{W_i \mid i = 1, \ldots, p\}$ is a cover of M, there exists i such that $x \in W_i$. If $y \notin \text{Cl } W_i$, then $h_i(y) < 1$ while $h_i(x) = 1$, in which case $e(x) \neq e(y)$. If $y \in \text{Cl } W_i$, then $\varphi_i(x) \neq \varphi_i(y)$, since φ_i is an embedding, so $\hat{\varphi}_i(x) \neq \hat{\varphi}_i(y)$, in which case $e(x) \neq e(y)$.

e is an immersion: Let $\bar{x} \in M$. We need to show that e has rank m at $\bar{x}$. Now there exists i such that $\bar{x} \in W_i$. We must show that $D(e\varphi_i^{-1})(\bar{x})$ has rank m, i.e., that this jacobian matrix has m linearly independent rows. Since $\hat{\varphi}_i | W_i = \varphi_i | W_i$, the m × m submatrix of $D(e\varphi_i^{-1})(\bar{x})$ consisting of those rows corresponding to the $\hat{\varphi}_i$ factor of e, i.e., rows

$$(i-1)(m+1) + 1, \ldots, (i-1)(m+1) + m$$

is the identity matrix, so these rows are linearly independent, as required.

Since e is an injective immersion having compact domain, e must be an embedding. □

According to Theorem 2, we can do all of our compact differential topology in euclidean space by studying only submanifolds of euclidean space. In particular, this means that it really *is* worthwhile drawing pictures to help our understanding of differential topology.

Two questions are prompted by Theorem 2. Do we need such a high dimensional euclidean space as $\mathbb{R}^{p(m+1)}$ in which to embed an m manifold? How big need p be anyway? Now p is the number of elements in a certain cover of a manifold by coordinate charts. One can show that by choosing the coordinate charts carefully we can take $p \leq m + 1$, so that we can embed any compact m manifold in $\mathbb{R}^{(m+1)^2}$. Using a different approach in Chap. 15, we will show how to embed any compact m manifold in $\mathbb{R}^{2m+1}$.

The second question is whether noncompact manifolds embed in euclidean space. Some do; $\mathbb{R}^m$ itself embeds in $\mathbb{R}^m$. Some don't, but this in turn leads to another question. Given two topological spaces X and Y, how can we decide that X does not embed in Y? The basic method is to find a topological property possessed by all subspaces of Y but not by X. Thus no non-Hausdorff space embeds in euclidean space. In the absence of a sufficiently large collection of topological properties, we will just give an example of a manifold which does not embed in any euclidean space.

Example Let Λ be an uncountable set and let M^m be any (nonempty) manifold. Consider

$$\Lambda \times M = \{(\lambda, x) \mid \lambda \in \Lambda, x \in M\}$$

For each $\lambda \in \Lambda$, define $i_\lambda : M \longrightarrow \Lambda \times M$ by $i_\lambda(x) = (\lambda, x)$. Decree a subset U of $\Lambda \times M$ to be open iff for all $\lambda \in \Lambda$, $i_\lambda^{-1}(U)$ is open in M. In effect, $\Lambda \times M$ consists of $\#(\Lambda)$ distinct copies of M on each of which may be imposed the differential structure of M. Then $\Lambda \times M$ does not embed in any euclidean space; there are too many copies of M to fit. (The topological property we are using here is the one which says that every uncountable subset has a limit point; this property holds in euclidean space but not in $\Lambda \times M$.) □

7. Submanifolds and an Embedding Theorem

The above example is unsatisfactory in the sense that it is not connected. One can construct a manifold (in our sense) which is connected, yet does not embed in euclidean space. Even so, there is a general result concerning embeddability of noncompact manifolds in euclidean space: every σ-compact manifold embeds in euclidean space. (σ-compact means "countable union of compact sets".)

EXERCISES

1. Prove that if M is a submanifold of N and N is a submanifold of P then M is a submanifold of P.
2. Prove that T^2 is a submanifold of $\mathbb{R}^3$.
3. Prove that S^m is a submanifold of S^n ($m \leq n$).
4. Prove that P^m is a submanifold of P^n ($m \leq n$).
5. Show how to modify the proof of Theorem 1 to obtain the result stated in the remark immediately following the theorem.
6. Let $(M^m, \mathcal{D})$ and (N^n, E) be two C^r manifolds and $f : M \to N$ a function. Let $M \times N = \{(x,y) \mid x \in M, y \in N\}$, topologized by the product topology, and $\Gamma(f) = \{(x,y) \in M \times N \mid y = f(x)\}$. Show that $\{(U \times V, \varphi \times \psi) \mid (U, \varphi) \in \mathcal{D}, (V, \psi) \in E\}$ is a basis for a differential structure $\mathcal{D} \times E$ on $M \times N$. Show that if $\gamma : M \to \Gamma(f)$ is defined by $\gamma(x) = (x, f(x))$, then $\gamma(\mathcal{D}) = \{(\gamma(U), \varphi\gamma^{-1}) \mid (U, \varphi) \in \mathcal{D}\}$ is a differential structure on $\Gamma(f)$. Prove that $\Gamma(f)$ is a submanifold of $M \times N$ if and only if f is differentiable.

8
TANGENT SPACES

In one-dimensional calculus, the derivative is closely allied to the tangent line to a curve. Similarly, in higher dimensions we can talk about the tangent hyperplane to a hypersurface and this, too, is related to the derivative. We should not be surprised, then, to find tangent spaces cropping up in our study of differentiable manifolds. A tangent vector will measure the way in which a given real-valued function with domain a differentiable manifold changes in a particular direction at a point on the manifold, much like the partial derivatives of calculus. At some points the function will not change in any direction; such points will be critical points. We will find that away from critical points the point preimages discussed in the introduction to Chap. 4 are unchanged. A collection of tangent vectors, one at each point of a manifold will form, in Chap. 10, a vector field. A vector field naturally gives rise to a differential equation on the manifold, and integration of this equation enables us to study how adjacent point preimages fit together. One of the important uses of vector fields is in the study of dynamics. Integration of the resulting differential equation enables one to determine the path of a particle in our dynamical system.

There are several approaches to tangent spaces of differentiable manifolds (and one can even consider tangent spaces of nondifferentiable manifolds!). In this chapter we consider three approaches and relate them to one another. The first approach is rather neat, although only by relating it to the others do we get a good geometric picture of it.

If M and N are two C^r manifolds, where $r = 1, 2,\ldots, \infty$, let $C^r(M,N)$ denote the set of all C^r maps from M to N. We will take $r = \infty$, but much of the work is valid for $r < \infty$.

Consider the process of differentiation of a function from $\mathbb{R}$ to $\mathbb{R}$. This process assigns to a differentiable function another function. Thus differentiation itself is a function with domain $C^1(\mathbb{R}, \mathbb{R})$ and range the continuous functions from $\mathbb{R}$ to $\mathbb{R}$. Restricting to $C^\infty(\mathbb{R}, \mathbb{R})$, we get a function $D : C^\infty(\mathbb{R}, \mathbb{R}) \rightarrow C^\infty(\mathbb{R}, \mathbb{R})$ defined by $D(f) = f'$. Consider now differentiation at a point $p \in \mathbb{R}$; call it D_p. Then $D_p : C^\infty(\mathbb{R}, \mathbb{R}) \rightarrow \mathbb{R}$ is given by $D_p(f) = f'(p) = D(f)(p)$, the jacobian matrix of Chap. 4. This function satisfies the familiar rules of elementary calculus: linearity, Leibnitz' formula, and localization [i.e., $D_p(f)$ depends only on the behavior of f in a neighborhood of p rather than in the whole of $\mathbb{R}$]. This idea extends to the directional derivative of advanced calculus. Given $f \in C^\infty(\mathbb{R}^m, \mathbb{R})$ and $p \in \mathbb{R}^m$, we can define the directional derivative of f at p in the direction $v \in \mathbb{R}^m$ to be

$$D_v(f)(p) = \lim_{h \to 0} \frac{f(p + hv) - f(p)}{h}$$

We then obtain functions $D_v : C^\infty(\mathbb{R}^m, \mathbb{R}) \rightarrow C^\infty(\mathbb{R}^m, \mathbb{R})$ and $D_{v,p} : C^\infty(\mathbb{R}^m, \mathbb{R}) \rightarrow \mathbb{R}$ with $D_{v,p}$ satisfying the above three conditions. It is these three conditions which we abstract, in effect identifying differentiation in the direction v with the direction v itself. Because of localization, we can again replace $\mathbb{R}^m$ by M^m since they are locally the same.

Definition Let $p \in M^m$, where M is a manifold. By a *tangent vector at p* we mean a function $v : C^\infty(M, \mathbb{R}) \rightarrow \mathbb{R}$ satisfying:

Tang 1. For all $\alpha, \beta \in \mathbb{R}$ and all $f, g \in C^\infty(M, \mathbb{R})$, $v(\alpha f + \beta g) = \alpha v(f) + \beta v(g)$.

Tang 2. For all $f, g \in C^\infty(M, \mathbb{R})$, $v(f \times g) = v(f)g(p) + f(p)v(g)$.

Tang 3. For all $f, g \in C^\infty(M, \mathbb{R})$, if $f|U = g|U$ for some neighborhood U of p, then $v(f) = v(g)$.

8. Tangent Spaces

Note that the addition of Tang 1 and the multiplication of Tang 2 are of functions, not of real numbers. Thus for $\alpha, \beta \in \mathbb{R}$ and $f, g \in C^\infty(M, \mathbb{R})$, the functions $\alpha f + \beta g$ and $f \times g$, both also members of $C^\infty(M, \mathbb{R})$, are defined by the equations

$$(\alpha f + \beta g)(x) = \alpha f(x) + \beta g(x)$$

$$(f \times g)(x) = f(x) g(x)$$

for each $x \in M$. The addition and multiplication on the right hand sides of these equations are those of real numbers, since all of the quantities involved are real numbers.

As we might hope for a differentiation operator, if v is a tangent vector at p and $f \in C^\infty(M, \mathbb{R})$ is constant, then $v(f) = 0$. Indeed choose $g \in C^\infty(M, \mathbb{R})$ so that $g(p) \neq 0$; for example, let $g(x) = 1$ for all $x \in M$. By Tang 2,

$$v(f \times g) = v(f) g(p) + f(p) v(g)$$

On the other hand, since f is constant, the product function $f \times g$ is the same as the scalar multiplication function $f(p)g$. Applying Tang 1 with $\alpha = 0$ and $\beta = f(p)$, we get

$$v(f \times g) = v(f(p)g) = f(p) v(g)$$

Thus $v(f) g(p) = 0$, so that $v(f) = 0$ as claimed.

Let TM_p denote the set of all tangent vectors at p. Then TM_p is called the *tangent space* to M at p.

Example The simplest (albeit trivial) example of a tangent vector is the zero vector, i.e., it assigns to each $f \in C^\infty(M, \mathbb{R})$ the real number 0. The directional derivatives are all tangent vectors (with $M = \mathbb{R}^m$). In fact, by Theorem 1 below, directional derivatives are the only tangent vectors to $\mathbb{R}^m$.

Now suppose M^m is an arbitrary manifold and (U, φ) is a chart about $p \in M$. Define $(\partial/\partial \varphi_i)|_p : C^\infty(M, \mathbb{R}) \to \mathbb{R}$ by assigning to $f \in C^\infty(M, \mathbb{R})$ the real number $\partial (f \varphi^{-1})/\partial x_i |_{\varphi(p)}$. Then $(\partial/\partial \varphi_i)|_p \in TM_p$.

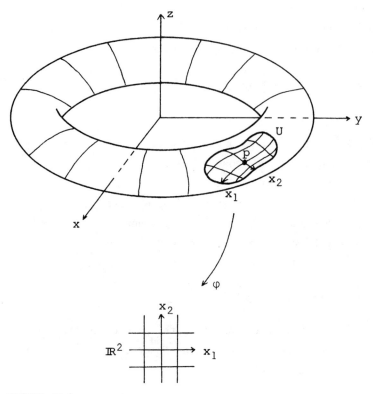

FIGURE 34

In effect, φ transfers the coordinate system of $\mathbb{R}^m$ to U; $(\partial/\partial\varphi_i)|_p$ is the derivative of f in the direction of the ith axis. Thus with (U,φ) as in Fig. 34, if $f,g,h \in C^\infty(T^2,\mathbb{R})$ denote, respectively, projection on the x, y, and z coordinates, then $(\partial f/\partial\varphi_1)|_p$ and $(\partial g/\partial\varphi_2)|_p$ are positive, $(\partial f/\partial\varphi_2)|_p$ and $(\partial g/\partial\varphi_1)|_p$ are about 0 and $(\partial h/\partial\varphi_1)|_p$ and $(\partial h/\partial\varphi_2)|_p$ are negative.

THEOREM 1. Let M^m be a manifold and $p \in M$. Then TM_p is an m dimensional vector space over $\mathbb{R}$.

Proof: Addition and scalar multiplication on TM_p are defined in the standard way by

8. Tangent Spaces

$$(u + v)(f) = u(f) + v(f) \qquad (rv)(f) = rv(f) \qquad f \in C^{\infty}(M, \mathbb{R})$$

Taking $X = C^{\infty}(M, \mathbb{R})$ in Exercise 1, we see that it suffices to show that TM_p is closed under these two operations, i.e., if $u, v \in TM_p$ and $r \in \mathbb{R}$, then $u + v \in TM_p$ and $rv \in TM_p$. We will verify that $u + v$ satisfies Tang 1, leaving the others as an exercise. Let $\alpha, \beta \in \mathbb{R}$ and $f, g \in C^{\infty}(M, \mathbb{R})$. Then

$$\begin{aligned}
(u + v)(\alpha f + \beta g) &= u(\alpha f + \beta g) + v(\alpha f + \beta g) & &\text{definition of vector addition} \\
&= \alpha u(f) + \beta u(g) + \alpha v(f) + \beta v(g) & &\text{u and v satisfy Tang 1} \\
&= \alpha[u(f) + v(f)] + \beta[u(g) + v(g)] & &\text{axioms for } \mathbb{R} \\
&= \alpha(u + v)(f) + \beta(u + v)(g) & &\text{definition of vector addition}
\end{aligned}$$

Thus $u + v$ satisfies Tang 1.

To verify that TM_p has dimension m, we must exhibit a basis having m elements. We use the advanced calculus result that the directional derivative is a linear combination of the partial derivatives to guide us, using the vectors $(\partial/\partial \varphi_i)\big|_p$ in the example above.

Using Lemma 5.1, we may choose a chart (U, φ) about p so that $\varphi(p) = 0$. We will show that

$$\left\{ \frac{\partial}{\partial \varphi_i}\bigg|_p \;\bigg|\; i = 1, \ldots, m \right\}$$

is a basis for TM_p. Thus we must show that these vectors span TM_p and that they are linearly independent.

Let $\varphi_i : U \to \mathbb{R}$ be defined by $\varphi_i(q) =$ the ith coordinate of $\varphi(q)$. As in the proof of Theorem 7.2, we may use Lemma 4.1 to extend over M the restriction of φ_i to a smaller neighborhood of p; call such an extension φ_i also. Thus $\varphi_i : M \to \mathbb{R}$ is a C^{∞} function such that for all q in some neighborhood of p, $\varphi_i(q)$ is the ith coordinate of $\varphi(q)$.

To show that the alleged basis spans TM_p, suppose that $v \in TM_p$. Let $\alpha_i = v(\varphi_i)$. By Tang 3, α_i is independent of the way φ_i was extended. The number $v(\varphi_i)$ is rather like the scalar product of two vectors, one being v and the other a "unit" vector in the direction of the ith coordinate axis. We claim that

$$v = \sum_{i=1}^{m} \alpha_i \left.\frac{\partial}{\partial \varphi_i}\right|_p$$

This is a sensible claim because if v were a vector in $\mathbb{R}^m$ then to express v as a linear combination of the standard basis vectors, to find the appropriate coefficients we would take the scalar product of v with the standard basis vectors.

To verify this claim we must show that, given $f \in C^\infty(M,\mathbb{R})$,

$$v(f) = \sum \alpha_i \left.\frac{\partial f}{\partial \varphi_i}\right|_p$$

Now, for x in a neighborhood of 0 in $\mathbb{R}^m$,

$$f\varphi^{-1}(x) = \int_0^1 \frac{d}{dt}\left[f\varphi^{-1}(tx)\right] dt + f(p)$$

$$= \int_0^1 \sum_{i=1}^{m} \left.\frac{\partial (f\varphi^{-1})}{\partial x_i}\right|_{tx} x_i \, dt + f(p) \qquad \text{chain rule}$$

$$= \sum_{i=1}^{m} x_i \int_0^1 \left.\frac{\partial f}{\partial \varphi_i}\right|_{\varphi^{-1}(tx)} dt + f(p)$$

Use Lemma 4.1 to find functions $g_i \in C^\infty(M,\mathbb{R})$ satisfying

$$g_i(q) = \int_0^1 \left.\frac{\partial f}{\partial \varphi_i}\right|_{\varphi^{-1}(t\varphi(q))} dt$$

8. Tangent Spaces

for q in some neighborhood of p. Then $g_i(p)$ should be the derivative of f in the ith coordinate direction and in fact, since $\varphi(p) = 0$, we have

$$g_i(p) = \int_0^1 \frac{\partial f}{\partial \varphi_i}\bigg|_{\varphi^{-1}(t,0)} dt = \frac{\partial f}{\partial \varphi_i}\bigg|_p \int_0^1 dt = \frac{\partial f}{\partial \varphi_i}\bigg|_p$$

Moreover, for q in a neighborhood of p, we have

$$f(q) = \sum_i \varphi_i(q) g_i(q) + f(p)$$

cf. the Taylor's expansion obtained in the proof of Theorem 4.5. Thus, in this neighborhood, f is a sum of products of functions together with a constant function. Thus

$$\begin{aligned}
v(f) &= v(\sum_i \varphi_i \times g_i + f(p)) && \text{by Tang 3} \\
&= \sum v(\varphi_i \times g_i) + 0 && \text{by Tang 1} \\
&= \sum [v(\varphi_i) g_i(p) + \varphi_i(p) v(g_i)] && \text{by Tang 2} \\
&= \sum \alpha_i \frac{\partial f}{\partial \varphi_i}\bigg|_p + 0
\end{aligned}$$

as required.

Linear independence of the vectors in the alleged basis follows from the facts that

$$\frac{\partial \varphi_i}{\partial \varphi_i}\bigg|_p = 1 \qquad \frac{\partial \varphi_j}{\partial \varphi_i}\bigg|_p = 0 \qquad \text{for } i \neq j$$

For if $\sum \alpha_i (\partial/\partial \varphi_i)\big|_p = 0$, then $\sum \alpha_i (\partial f/\partial \varphi_i)\big|_p = 0$ for all $f \in C^\infty(M, \mathbb{R})$. In particular, setting $f = \varphi_j$, we get

$$0 = \sum \alpha_i \left.\frac{\partial \varphi_j}{\partial \varphi_i}\right|_p = \alpha_j \quad \square$$

Definition The ordered m-tuple $(\alpha_1,\ldots,\alpha_m)$ of Theorem 1 is called the *components* of the vector v with respect to the chart (U,φ).

If (V,ψ) is another chart about p and $(\beta_1,\ldots,\beta_m)$ are the components of v with respect to (V,ψ), then

$$\beta_i = v(\psi_i) = \sum_{j=1}^m \alpha_j \left.\frac{\partial \psi_i}{\partial \varphi_j}\right|_p = \sum_{j=1}^m \alpha_j \left.\frac{\partial (\psi_i \varphi^{-1})}{\partial x_j}\right|_{\varphi(p)}$$

But $\left.\partial(\psi_i\varphi^{-1})/\partial x_j\right|_{\varphi(p)}$ is the (i,j) element of the jacobian matrix $D(\psi\varphi^{-1})$, so we have

$$\begin{pmatrix} \beta_1 \\ \vdots \\ \beta_m \end{pmatrix} = D(\psi\varphi^{-1})(\varphi(p)) \times \begin{pmatrix} \alpha_1 \\ \vdots \\ \alpha_m \end{pmatrix}$$

where the multiplication is matrix multiplication. This suggests a geometrically more satisfying definition of tangent vector—our second approach.

Definition Let $p \in M^m$, where $(M,\mathcal{D})$ is a manifold. By a *tangent vector at p* we mean a function v with domain

$$\{(U,\varphi) \in \mathcal{D} \mid U \text{ is a neighborhood of } p\}$$

and range the column matrices of size $m \times 1$ of real numbers such that if $(U,\varphi),(V,\psi) \in \mathcal{D}$ are both charts about p, then

$$v(V,\psi) = D(\psi\varphi^{-1})(\varphi(p)) \times v(U,\varphi)$$

8. Tangent Spaces

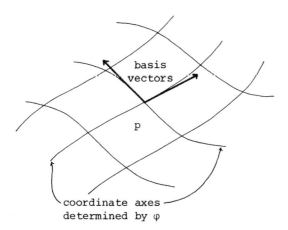

FIGURE 35

The geometric interpretation of this definition is as follows (see Fig. 35). Through the point p, construct coordinate axes determined by the chart (U,φ). The tangent lines to these coordinate axes form a basis for the tangent space. The column matrix $v(U,\varphi)$ determines the coordinates of the vector v with respect to this basis, and the equation relating $v(V,\psi)$ to $v(U,\varphi)$ shows how to change the coordinates from one basis to the other.

One can easily check that the function which assigns to any vector $v \in TM_p$ the function

$(U,\varphi) \longmapsto$ column matrix of components of v with respect to (U,φ)

is a vector space isomorphism, so that the first two approaches to tangent vectors are essentially the same.

The above geometric interpretation motivates the third approach to tangent vectors.

Definition Let M be a manifold. A *curve* on M is a differentiable function $\gamma : I \longrightarrow M$, where I is an interval in $\mathbb{R}$. Given a curve γ we may define a tangent vector v to M at $\gamma(t_0)$, $t_0 \in I$, by setting

$$v(f) = (f\gamma)'(t_0) \qquad \text{for all } f \in C^\infty(M, \mathbb{R})$$

Note that $f\gamma : I \to \mathbb{R}$; $(f\gamma)'$ denotes the ordinary derivative of $f\gamma$. The vector v just defined is called the *velocity vector* of γ at t_0.

On the other hand, given $v \in TM_p$, v is the velocity vector of some curve. Indeed, let (U,φ) be any chart with $\varphi(p) = 0$, and let $(\alpha_1, \ldots, \alpha_m)$ be the components of v with respect to (U,φ). As illustrated by Fig. 36, for t near 0, define

$$\gamma(t) = \varphi^{-1}(\alpha_1 t, \ldots, \alpha_m t)$$

Then for all $f \in C^\infty(M, \mathbb{R})$,

$$(f\gamma)'(0) = v(f) \qquad \gamma(0) = p$$

In the particular case where M is a submanifold of $\mathbb{R}^n$, we can use the above ideas to obtain a natural identification of TM_p as an affine subspace of the vector space $\mathbb{R}^n$. An affine subspace of $\mathbb{R}^n$ is a translation of a vector subspace of $\mathbb{R}^n$, i.e., a set of the form

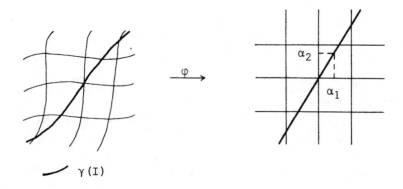

FIGURE 36

8. *Tangent Spaces*

$$p + V = \{p + v \mid v \in V\}$$

where $p \in \mathbb{R}^n$ is a fixed point and V is a vector subspace of $\mathbb{R}^n$. Thus if V is of dimension 1, i.e., a straight line through 0, then p + V is a parallel straight line through the point p. If V is of dimension 2, then p + V is a parallel plane through p, etc.

PROPOSITION 2. Let M^m be a submanifold of $\mathbb{R}^n$ and let $p \in M$. Then there is an m-dimensional affine subspace of $\mathbb{R}^n$ containing p which also contains the tangent line at p of every curve in M passing through p.

Proof: Since M is a submanifold of $\mathbb{R}^n$, we may choose a chart (U,φ) on $\mathbb{R}^n$ about p such that $\varphi^{-1}(\mathbb{R}^m) = U \cap M$ and $(U \cap M, \varphi \mid U \cap M)$ is a chart in the structure of M. We may further assume (cf. Lemma 5.1) that $\varphi(p) = 0$. The idea of the proof is to choose m canonical directions based at p. These directions will actually be the directions of the first m coordinate axes of (U,φ), all of which lie in M. We will need to show that they are independent, but this should not be surprising. We will then show that the tangent line at p of every curve in M passing through p lies in the resulting affine space by decomposing the curve into a linear combination of curves along the first m coordinate axes of (U,φ).

Define $\delta_i : I \to M \subset \mathbb{R}^n$ (i = 1, ..., m) by $\delta_i(t) = \varphi^{-1}(0,...,0,t,0,...,0)$ where t is in the ith position, and I is some interval containing 0. Since φ is in the structure of $\mathbb{R}^n$, $\delta : I \to \mathbb{R}^n$ is a smooth function, hence differentiable at 0. It is claimed that the affine space through p parallel to the vector subspace of $\mathbb{R}^n$ spanned by $\{\delta_i'(0) \mid i = 1, ..., m\}$ is the required space. Letting T denote this space, we have

$$T = \left\{ p + \sum_{i=1}^{m} \alpha_i \delta_i'(0) \mid \alpha_i \in \mathbb{R} \text{ for each i} \right\}$$

First we show that T is an m-dimensional affine space. It is certainly an affine space. We need to check that its dimension is m, i.e., that the vectors $\{\delta_i'(0) \mid i = 1, \ldots, m\}$ are linearly independent. Since $\delta_i(t) = \varphi^{-1}(0,\ldots 0,t,0,\ldots,0)$, the chain rule tells us that

$$\delta_i'(0) = D(\varphi^{-1})(0) \begin{pmatrix} 0 \\ \vdots \\ 1 \\ \vdots \\ 0 \end{pmatrix}$$

where 1 is in the ith position in the column matrix. Since φ^{-1} is a diffeomorphism of a neighborhood of 0, $D(\varphi^{-1})(0)$ is a vector space isomorphism. In particular it carries the m linearly independent column matrices

$$\begin{pmatrix} 1 \\ 0 \\ \vdots \\ 0 \end{pmatrix}, \begin{pmatrix} 0 \\ 1 \\ 0 \\ \vdots \\ 0 \end{pmatrix} \cdots \begin{pmatrix} 0 \\ \vdots \\ 0 \\ 1 \\ 0 \\ \vdots \\ 0 \end{pmatrix}$$

into m linearly independent vectors.

Thus $\{\delta_i'(0) \mid i = 1, \ldots, m\}$ are m linearly independent vectors, so they span a vector subspace of $\mathbb{R}^n$ of dimension m. T, being a translate of this subspace, is thus an m-dimensional affine space.

Clearly $p \in T$, so it remains to show that T contains the tangent line at p of every curve in M passing through p. Suppose $\gamma : I \to M$ is such a curve, say $\gamma(t_0) = p$. The tangent line of γ at t_0 is given by

$$\{p + k\gamma'(t_0) \mid k \in \mathbb{R}\}$$

8. Tangent Spaces

Consider $\varphi\gamma$. Since $\varphi\gamma(I) \subset \mathbb{R}^m$, the vector $D(\varphi)(p)\gamma'(t_0)$ is a vector of $\mathbb{R}^m$, so is a linear combination of the standard basis, say

$$D(\varphi)(p)\gamma'(t_0) = \sum_{i=1}^{m} \alpha_i \begin{pmatrix} 0 \\ \vdots \\ 1 \\ \vdots \\ 0 \end{pmatrix}$$

Therefore

$$\gamma'(t_0) = \sum_{i=1}^{m} \alpha_i [D(\varphi)(p)]^{-1} \begin{pmatrix} 0 \\ \vdots \\ 1 \\ \vdots \\ 0 \end{pmatrix}$$

$$= \sum_{i=1}^{m} \alpha_i \, D(\varphi^{-1})(0) \begin{pmatrix} 0 \\ \vdots \\ 1 \\ \vdots \\ 0 \end{pmatrix}$$

$$= \sum_{i=1}^{m} \alpha_i \delta_i'(0)$$

Therefore

$$p + k\gamma'(t_0) = p + \sum_{i=1}^{m} k\alpha_i \delta_i'(0) \in T$$

Thus T contains all of the tangent lines. □

Any (n-1)-dimensional affine subspace of $\mathbb{R}^n$ containing the space T of Proposition 2 is called a *tangent hyperplane* to M at p in $\mathbb{R}^n$.

EXERCISES

1. Let X be any set and V be a vector space over $\mathbb{R}$. Let $F(X,V) = \{f : X \to V\}$, and define addition and scalar multiplication on $F(X,V)$ by

$$(f + g)(x) = f(x) + g(x)$$
$$(rf)(x) = rf(x)$$

Verify that with these operations, $F(X,V)$ is a vector space over $\mathbb{R}$.

2. Complete the proof that TM_p is a vector space over $\mathbb{R}$.

3. Consider the basis for the structure of S^2 described in Chap. 6 (taking m = 2). Let v be the tangent vector to S^2 at (0,1,0) whose components with respect to (U,φ) are (1,0). What are the components of v with respect to (V,ψ)? Illustrate on a diagram a curve for which v is the velocity vector at (0,1,0).

4. Suppose $f : M \to N$ is differentiable and let $p \in M$. Define $df_p : TM_p \to TN_{f(p)}$ by

$$df_p(v)(g) = v(gf) \qquad \text{for all } v \in TM_p, \text{ for all } g \in C^\infty(N,\mathbb{R})$$

Show that df_p is a linear transformation and that the rank of f at p is the same as the rank of the linear transformation df_p. This exercise gives us a way of transferring the jacobian to a manifold.

5. Let $f : M \to N$ be an immersion. According to Exercise 4, for all $p \in M$, $df_p(TM_p)$ is a vector subspace of $TN_{f(p)}$. Consider a specific immersion of S^1 as a figure eight in S^2. Illustrate how $df_p(TS^1_p)$ sits inside the affine subspace of $\mathbb{R}^3$ obtained by applying Proposition 2 to S^2 at $f(p)$. How is this set related to curves on S^1 at p? How does this latter result generalize to the case where f is any immersion with $N \subset \mathbb{R}^q$ (some q)?

9
CRITICAL POINTS AGAIN

In this chapter, the notions of critical point and index from Chap. 4 are transferred to manifolds. We also introduce two important illustrative examples to which we will refer repeatedly. The chapter ends with the definition of a Morse function and the statement (but not its proof) that every compact manifold supports a Morse function. Morse functions will be our basic tool in the classification of surfaces. Corollary 2 of this chapter tells us that the preimage under a Morse function of a regular value is a submanifold of prescribed dimension. Theorem 2 of Chap. 10 carries this a step further to show how adjacent preimages are related when there are no critical points nearby. It turns out that the relationship is rather trivial. On the other hand, preimages of critical values are much more complicated, and their study takes up Chaps 11, 12, and 13. Initially it might seem hopeless to classify all possibilities, but Morse's theorem, which is the manifold analogue of Theorem 4.5, provides us with a considerable simplification.

Definition Let $f : M^m \to N^n$ be differentiable. Then $p \in M$ is a *regular point* of f iff the rank of f at p is n. Otherwise, p is a *critical point*. A point $q \in N$ is a *regular value* of f iff every point of $f^{-1}(q)$ is regular; otherwise q is a *critical value*. Note that if $q \in N - f(M)$ then $f^{-1}(q) = \phi$, so q is a regular value. If $m < n$, every point of M is critical, since the rank can never exceed m. If all of the points of the domain of f are regular then f is called a *submersion* (cf. immersion introduced in Chap. 5). We will not have any use for submersions.

The following two examples are standard in the sense that we will use such functions in Chap. 14 to classify surfaces. In Chap. 13, we will find that there are certain basic changes in the shape of a surface. The function f below illustrates all four possibilities in the case of an orientable surface. Precise details concerning these points will gradually emerge in the intervening chapters.

Examples Recall T^2, obtained by revolving the circle $(x - 2)^2 + z^2 = 1$ in the xz plane about the z axis. Define $f, g : T^2 \to \mathbb{R}$ by $f(x,y,z) = x$ and $g(x,y,z) = z$.

Using the basis for the structure of T^2 described in Chap. 5, we now locate the critical points of f and g. By reference to Fig. 19, we should expect to find critical points of f precisely where the x axis pierces T^2, i.e., at $(\pm 3, 0, 0)$ and $(\pm 1, 0, 0)$, and critical points of g precisely on the uppermost and lowermost circles, i.e., the $\pi/2$ and $3\pi/2$ lines of longitude.

To avoid confusion, we will use (a,b) to denote the coordinates of a point of the range of φ and ψ, i.e., of $\mathbb{R}^2$.

To locate the critical points of f and g we must calculate the ranks of $f\varphi^{-1}$, $f\psi^{-1}$, $g\varphi^{-1}$, and $g\psi^{-1}$. This entails determining φ^{-1} and ψ^{-1}. In fact we determine an immersion (= submersion, since domain and range have the same dimension) $\chi : \mathbb{R}^2 - \{0\} \to T^2$ satisfying $\chi|\varphi(U) = \varphi^{-1}$ and $\chi|\psi(V) = \psi^{-1}$ (from which the immersivity of χ follows immediately; cf. Exercise 5.7). It will then suffice to calculate the ranks of $f\chi$ and $g\chi$ on $\varphi(U) \cup \psi(V)$.

Just as φ^{-1} and ψ^{-1} wrap annuli of $\mathbb{R}^2$ around T^2, the map χ wraps $\mathbb{R}^2 - \{0\}$ around T^2 infinitely many times, any circle centered at 0 being mapped to a line of longitude, and any annulus between two such circles whose radii differ by 2π being wrapped around T^2 once. Precisely,

$$\chi(a,b) = \left(\frac{a}{\sqrt{a^2+b^2}} (2 + \cos\sqrt{a^2+b^2}), \frac{b}{\sqrt{a^2+b^2}} (2 + \cos\sqrt{a^2+b^2}), \sin\sqrt{a^2+b^2} \right)$$

9. Critical Points Again

Consider the function $f\chi : \mathbb{R}^2 - \{0\} \to \mathbb{R}$. We have

$$f\chi(a,b) = \frac{a}{\sqrt{a^2+b^2}}(2 + \cos\sqrt{a^2+b^2})$$

so

$Df\chi(a,b)^*$

$$= \begin{bmatrix} \dfrac{1}{\sqrt{a^2+b^2}}(2+\cos\sqrt{a^2+b^2}) - \dfrac{a^2}{(a^2+b^2)^{3/2}}(2+\cos\sqrt{a^2+b^2}) - \dfrac{a^2 \sin\sqrt{a^2+b^2}}{a^2+b^2} \\[2ex] -\dfrac{ab}{(a^2+b^2)^{3/2}}(2+\cos\sqrt{a^2+b^2}) - \dfrac{ab\sin\sqrt{a^2+b^2}}{a^2+b^2} \end{bmatrix}$$

$$= \begin{bmatrix} \dfrac{b^2(2+\cos\sqrt{a^2+b^2})}{(a^2+b^2)^{3/2}} - \dfrac{a^2\sin\sqrt{a^2+b^2}}{a^2+b^2} \\[2ex] -\dfrac{ab}{(a^2+b^2)^{3/2}}(2+\cos\sqrt{a^2+b^2} + \sqrt{a^2+b^2}\sin\sqrt{a^2+b^2}) \end{bmatrix}$$

where $Df\chi(a,b)^*$ is the transpose of the 1×2 matrix $Df\chi(a,b)$. Now $\chi(a,b)$ is a critical point of f iff f has rank < 1, hence 0, at $\chi(a,b)$. Thus $\chi(a,b)$ is a critical point of f provided both entries of $Df\chi(a,b)$ are 0, hence provided

$$b^2(2+\cos\sqrt{a^2+b^2}) = a^2\sqrt{a^2+b^2}\sin\sqrt{a^2+b^2} \qquad (1)$$

$$ab(2+\cos\sqrt{a^2+b^2} + \sqrt{a^2+b^2}\sin\sqrt{a^2+b^2}) = 0 \qquad (2)$$

From (2), there are three possibilities: i. $a = 0$; ii. $b = 0$; iii. $2 + \cos\sqrt{a^2+b^2} + \sqrt{a^2+b^2}\sin\sqrt{a^2+b^2} = 0$.

i. If $a = 0$, then (1) reduces to $b^2(2+\cos|b|) = 0$, so $b = 0$ since $2 + \cos|b| \geq 1$. We cannot have $a = 0$ and $b = 0$ since $(0,0)$ is not in the domain of χ. Thus case i does not arise.

ii. If $b = 0$, then (1) reduces to $0 = a^3 \sin |a|$. Again we cannot have $a = 0$, so $\sin |a| = 0$, hence $a = n\pi$ ($n = \pm 1, \pm 2, \ldots$).

iii. If $2 + \cos \sqrt{a^2 + b^2} + \sqrt{a^2 + b^2} \sin \sqrt{a^2 + b^2} = 0$, then

$$b^2(2 + \cos \sqrt{a^2 + b^2}) = -b^2 \sqrt{a^2 + b^2} \sin \sqrt{a^2 + b^2}$$

so substituting in (1) we get

$$(a^2 + b^2)^{3/2} \sin \sqrt{a^2 + b^2} = 0$$

and hence $\sin \sqrt{a^2 + b^2} = 0$. On substituting this in the above equation we find that $2 + \cos \sqrt{a^2 + b^2} = 0$, which is impossible since $2 + \cos \sqrt{a^2 + b^2} \geq 1$.

Thus the only critical points of f are $\chi(n\pi, 0)$, where n runs through all nonzero integers. Note that for such n, $(n\pi, 0) \in \varphi(U) \cup \psi(V)$ only if $n = \pm 1$ and ± 2. Evaluating, we get $\chi(\pm \pi, 0) = (\pm 1, 0, 0)$ and $\chi(\pm 2\pi, 0) = (\pm 3, 0, 0)$, which are the critical points of f predicted above.

Now consider the function $g\chi : \mathbb{R}^2 - \{0\} \longrightarrow \mathbb{R}$. We have

$$g\chi(a,b) = \sin \sqrt{a^2 + b^2}$$

so

$$Dg\chi(a,b) = \left(\frac{a \cos \sqrt{a^2 + b^2}}{\sqrt{a^2 + b^2}}, \frac{b \cos \sqrt{a^2 + b^2}}{\sqrt{a^2 + b^2}} \right)$$

Thus $\chi(a,b)$ is a critical point of g provided $\cos \sqrt{a^2 + b^2} = 0$. Now $(a,b) \in \varphi(U) \cup \psi(V)$ satisfies this equation only if (a,b) lies on one of the circles of radius $\pi/2$, $3\pi/2$, or $5\pi/2$, centered at $(0,0)$. These circles map under χ to the circles of critical points of g predicted above.

Note that in the calculations above, we have sought all critical points of f and g, and to enable us to be sure that we missed none,

9. Critical Points Again

we needed to consider sufficient charts to have a basis for the differential structure of T^2. If, however, we want merely to determine whether a particular point is a critical point, it might be better to use some other chart. For example, to check that $(3,0,0)$ is a critical point of f, it might be better to use the chart (W,ω), where W is a small open neighborhood of $(3,0,0)$ and $\omega(x,y,z) = (y,z)$. Intuitively one would expect that (W,ω) is in the structure of T^2, and it is easily verified that it is. Since $\omega(3,0,0) = (0,0)$, to find the rank of f at $(3,0,0)$, we need to find the rank of the jacobian of $f\omega^{-1}$ at $(0,0)$. Now $f(x,y,z) = x$, so $f\omega^{-1}(y,z)$ is obtained by solving the equation $(\sqrt{x^2+y^2} - 2)^2 + z^2 = 1$ of the torus for x in terms of y and z. A little algebra leads to

$$x = \sqrt{(2 + \sqrt{1 - z^2})^2 - y^2}$$

where we take the positive square root each time to ensure that x is about 3 (the three other choices give rise to charts about the other three critical points). Thus

$$f\omega^{-1}(y,z) = \sqrt{(2 + \sqrt{1 - z^2})^2 - y^2}$$

and so

$$D(f\omega^{-1})(y,z) = \left(-\frac{y}{x} \quad \frac{-z(2 + \sqrt{1 - z^2})}{x\sqrt{1 - z^2}} \right)$$

$$D(f\omega^{-1})(0,0) = (0 \quad 0)$$

Since this last matrix has rank $0 < 1$, the point $\omega^{-1}(0,0) = (3,0,0)$ is a critical point of f.

THEOREM 1. Let $f : M^m \to N^n$ be differentiable and let $p \in M$. Then p is a regular point of f iff for each chart (V,ψ) in the structure of N with $\psi f(p) = 0$, there is a chart (U,φ) in the structure of M so that:

i. $f(U) \subset V$.

ii. $\varphi(p) = 0$.

iii. $\psi f \varphi^{-1}(x_1,\ldots,x_m) = (x_1,\ldots,x_n)$ for all $(x_1,\ldots,x_m) \in \varphi(U)$.

Proof: Suppose p is regular and (V,ψ) is a chart about $f(p)$ with $\psi f(p) = 0$. Let (U',φ') be any chart about p with $\varphi'(p) = 0$ and $f(U') \subset V$. By applying Corollary 4.4 to $\psi f \varphi'^{-1}$, we find a diffeomorphism h of a neighborhood of 0 in $\mathbb{R}^m$ onto another such neighborhood so that $h(0) = 0$ and

$$\psi f \varphi'^{-1} h(x_1,\ldots,x_m) = (x_1,\ldots,x_n)$$

Let $U = \varphi'^{-1}$ (domain of h) and $\varphi = h^{-1}\varphi'$. By DS2, (U,φ) is also in the structure of M. Further, (U,φ) satisfies the three conditions i, ii, and iii.

Conversely, if there is a pair of charts (V,ψ) and (U,φ) as above, then $\psi f \varphi^{-1}$ is clearly of rank n at 0, so f is of rank n at p, i.e., p is a regular point of f. □

Theorem 1 tells us that by judiciously choosing charts around a regular point, the function, in a neighborhood of the regular point, is just projection on the first n coordinates. This is about the nicest kind of function. As an application, we have the following important result.

COROLLARY 2. Let $f : M^m \to N^n$ be differentiable and let q be a regular value of f. Then $f^{-1}(q)$ has a natural differential structure which makes it into an (m-n)-submanifold of M.

Proof: If $q \notin f(M)$ there is nothing to prove for then $f^{-1}(q) = \phi$. Assume then that $q \in f(M)$ and let (V,ψ) be any chart about q with $\psi(q) = 0$. For each $p \in f^{-1}(q)$, let (U_p,φ_p) be a chart as given by Theorem 1. Comparing condition iii of Theorem 1 with the requirement $\varphi_p^{-1}(\mathbb{R}^{m-n}) = U_p \cap f^{-1}(q)$ for a submanifold, we see that instead of having

$$U_p \cap f^{-1}(q) = \varphi_p^{-1}\{(x_i) \in \mathbb{R}^m \mid x_j = 0 \text{ for all } j = m-n+1, \ldots, m\}$$

9. Critical Points Again

we have

$$U_p \cap f^{-1}(q) = \varphi_p^{-1}\{(x_i) \in \mathbb{R}^m \mid x_j = 0 \text{ for all } j = 1, \ldots, n\}$$

This situation is rectified by interchanging the first n and last m - n coordinates of $\mathbb{R}^m$. Define $\alpha : \mathbb{R}^m \to \mathbb{R}^m$ by

$$\alpha(x_1, \ldots, x_m) = (x_{n+1}, \ldots, x_m, x_1, \ldots, x_n)$$

Since α is a diffeomorphism, by DS2 $(U_p, \alpha\varphi_p)$ is also a chart in the structure of M. Furthermore,

$$(\alpha\varphi_p)^{-1}(\mathbb{R}^{m-n}) = U_p \cap f^{-1}(q)$$

In addition to being a requirement for a submanifold, this last equation enables us to define a differential structure on $f^{-1}(q)$ by specifying a basis. Precisely,

$$B_q = \{(U_p \cap f^{-1}(q), \alpha\varphi_p \mid U_p \cap f^{-1}(q)) \mid p \in f^{-1}(q)\}$$

is a basis for a differential structure on $f^{-1}(q)$. Clearly the open sets $U_p \cap f^{-1}(q)$ cover $f^{-1}(q)$, and if p_1 and p_2 are two points of $f^{-1}(q)$, then

$$(\alpha\varphi_{p_2} \mid U_{p_2} \cap f^{-1}(q))(\alpha\varphi_{p_1} \mid U_{p_1} \cap f^{-1}(q))^{-1}$$

is the restriction to $\mathbb{R}^{m-n}$ of $\alpha\varphi_{p_2}\varphi_{p_1}^{-1}\alpha^{-1}$. Since (U_{p_1}, φ_{p_1}) and (U_{p_2}, φ_{p_2}) are in the structure of M, $\varphi_{p_2}\varphi_{p_1}^{-1}$ is differentiable, so the coordinate transformations in B_q are differentiable. Thus B_q is a basis for a differential structure on $f^{-1}(q)$.

For each $p \in f^{-1}(q)$ the chart $(U_p, \alpha\varphi_p)$ satisfies the requirements of the definition of submanifold. □

Theorem 1 and Corollary 2 tell us that regular points and values are very well behaved and that there is not very much involved in

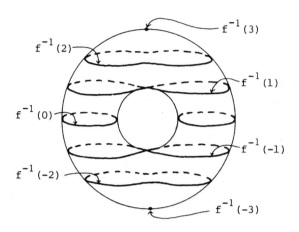

FIGURE 37

their study. Critical points, on the other hand, are much more complicated. In particular, Corollary 2 is not at all true for critical points.

Example Let $f,g : T^2 \to \mathbb{R}$ be as in the example above. Then f has four critical points and four critical values, viz., $\pm 1, \pm 3$. As Fig. 37 shows, if $|q| < 1$, then $f^{-1}(q)$ consists of two circles (up to diffeomorphism!), and if $1 < |q| < 3$ then $f^{-1}(q)$ consists of one circle, in each case a $(2 - 1 = 1)$-submanifold of T^2 as predicted by Corollary 2. $f^{-1}(\pm 3)$ is a point, a 0-submanifold of T^2, and $f^{-1}(\pm 1)$ is a figure eight, which is not a submanifold.

The function g has two critical values, viz., ± 1. If $|q| < 1$ then $g^{-1}(q)$ consists of two circles, a 1-submanifold of T^2, and $g^{-1}(\pm 1)$ consists of one circle, again a 1-submanifold of T^2. Note that every point of $g^{-1}(\pm 1)$ is a critical point, but this is not true in the case of $f^{-1}(\pm 1)$.

The above examples show that the inverse image of a critical value may be a submanifold of the domain of any dimension (the constant functions $T^2 \to \mathbb{R}$ illustrate that this set could be a 2-submanifold) or not even a manifold at all. It is by studying

9. Critical Points Again

the change in the character of the inverse image of a value as we cross a critical value that we will classify surfaces.

In a sense, all of the critical points of the function f above are necessary, whereas those of g are not. More precisely, any differentiable function $T^2 \to \mathbb{R}$ near f must have critical points near those of f (and of the same quality: one maximum, one minimum, two saddle points), but this is not true of g. The difference lies in the second derivative, which we will now exploit. Our attention will be restricted to functions whose range is $\mathbb{R}$ — the functions may then be thought of as height functions, assigning to each point its height above a given level (cf. the function $S^2 \to \mathbb{R}$ which assigns to each point of the earth's surface its height above sea level). The geographical analogy may be carried further, as indicated by the following theorem.

THEOREM 3. Let $f : M \to \mathbb{R}$ be differentiable with M compact. Then there is an embedding $e : M \to \mathbb{R}^n$, for some n, so that

i. For all $p \in M$, f(p) = nth coordinate of e(p).
ii. p is a critical point of f iff $x_n = f(p)$ is the equation of a tangent hyperplane to e(M) at e(p) in $\mathbb{R}^n$.

Proof: By Theorem 7.2, there is an embedding $e' : M \to \mathbb{R}^{n-1}$ for n sufficiently large. Then $e : M \to \mathbb{R}^n$ defined by letting the ith coordinate of e(p) be the ith coordinate of e'(p) for $i < n$, and the nth coordinate of e(p) be f(p), is an embedding satisfying condition i. We show that it satisfies ii as well.

There are two parts to the proof of ii. Suppose p is a critical point of f and let γ be a curve on e(M) through e(p), say $\gamma(t_0) = e(p)$. We need to show that the tangent line to γ at t_0 lies in the hyperplane $x_n = f(p)$, i.e., that the nth coordinate of the tangent line to γ at t_0 is constantly f(p). As in Proposition 8.2, the tangent line to γ at t_0 is

$$\{e(p) + k\gamma'(t_0) \mid k \in \mathbb{R}\}$$

In particular, then, the nth coordinate of any point on the tangent line is of the form $f(p) + k\gamma_n'(t_0)$, where γ_n is the nth coordinate of γ, and k ranges through $\mathbb{R}$. Thus we must show that $\gamma_n'(t_c) = 0$. Now $\gamma_n = fe^{-1}\gamma$, so if (U,φ) is a chart in the structure of M about p, then $\gamma_n = (f\varphi^{-1})(\varphi e^{-1}\gamma)$, so

$$\gamma_n'(t_0) = D(\gamma_n)(t_0) = D(f\varphi^{-1})(\varphi(p)) \times D(\varphi e^{-1}\gamma)(t_0)$$

But p is a critical point of f, so f has rank < 1 at p, i.e., f has rank 0 at p, so the matrix $D(f\varphi^{-1})(\varphi(p))$ is the zero matrix. Thus $\gamma_n'(t_0) = 0$ as required.

To prove the converse part of ii, suppose $x_n = f(p)$ determines a tangent hyperplane to $e(M)$ at $e(p)$ in $\mathbb{R}^n$. We must show that f has rank 0 at p, i.e., if (U,φ) is a chart on M with $\varphi(p) = 0$, then $D(f\varphi^{-1})(\varphi(p))$ has rank 0, thus is a matrix of zeros. Thus we must show that for all $i = 1, \ldots, m$, $(\partial f/\partial \varphi_i)|_p = 0$. Let $\delta_i : I \to M$ be the standard curve in the proof of Proposition 8.2, i.e., $\delta_i(t) = \varphi^{-1}(0,\ldots,0,t,0,\ldots,0)$, where t is in the ith position. The hyperplane $x_n = f(p)$ contains the tangent line to the curve $e\delta_i$ at 0; thus $(f\delta_i)'(0) = 0$. Since $(\partial f/\partial \varphi_i)|_p = (f\delta_i)'(0)$, we obtain the required result. □

Definition Let $f : M^m \to \mathbb{R}$ be differentiable and let $p \in M$. The *hessian* of f at p with respect to a chart (U,φ) about p is the m × m matrix whose (i,j) entry is $\partial^2(f\varphi^{-1})/\partial x_i \partial x_j |_{\varphi(p)}$. If p is a critical point of f, say that it is *nondegenerate* iff the hessian of f at p with respect to some chart about p is nonsingular. Otherwise call p a *degenerate* critical point. It is irrelevant which chart about p is chosen to check nondegeneracy; cf. exercise 4.6.

Examples The four critical points of the function $f : T^2 \to \mathbb{R}$ considered above are all nondegenerate, whereas those of $g : T^2 \to \mathbb{R}$ are all degenerate. We can verify these by using the above calculations of the jacobian to help us find the hessian with respect to the particular charts. For example,

9. Critical Points Again

$$Dg\chi(a,b) = \left(\frac{a \cos \sqrt{a^2+b^2}}{\sqrt{a^2+b^2}}, \frac{b \cos \sqrt{a^2+b^2}}{\sqrt{a^2+b^2}} \right)$$

with the critical points occurring when $\cos \sqrt{a^2+b^2} = 0$. Differentiating the above and setting $\cos \sqrt{a^2+b^2} = 0$ after the differentiation, we get

$$Hg\chi(a,b) \Big|_{\cos \sqrt{a^2+b^2}=0} = \begin{pmatrix} -\dfrac{a^2 \sin \sqrt{a^2+b^2}}{a^2+b^2} & -\dfrac{ab \sin \sqrt{a^2+b^2}}{a^2+b^2} \\ -\dfrac{ab \sin \sqrt{a^2+b^2}}{a^2+b^2} & -\dfrac{b^2 \sin \sqrt{a^2+b^2}}{a^2+b^2} \end{pmatrix}$$

$$= -\frac{\sin \sqrt{a^2+b^2}}{a^2+b^2} \begin{pmatrix} a^2 & ab \\ ab & b^2 \end{pmatrix}$$

which has determinant 0, hence is singular.

THEOREM 4 (Morse's Theorem). Let p be a nondegenerate critical point of the function $f : M^m \to \mathbb{R}$. Then there is a chart (U, φ) about p and a nonnegative integer λ such that $\varphi(p) = 0$ and for all $x = (x_1, \ldots, x_m) \in \varphi(U)$,

$$f\varphi^{-1}(x) = -\sum_{i=1}^{\lambda} x_i^2 + \sum_{i=\lambda+1}^{m} x_i^2 + f(p)$$

Proof: Let (V, ψ) be any chart about p with $\psi(p) = 0$. Consider the function $\bar{f} : \psi(V) \to \mathbb{R}$ given by

$$\bar{f}(x) = f\psi^{-1}(x) - f(p)$$

Then $\bar{f}$ is differentiable, $\bar{f}(0) = 0$, and 0 is a nondegenerate critical point of $\bar{f}$. Thus by Theorem 4.5 there is a diffeomorphism $\theta : V' \to \theta(V') \subset \mathbb{R}^m$, where V' and $\theta(V')$ are neighborhoods of 0 in $\mathbb{R}^m$ such that $\theta(0) = 0$ and

$$\bar{f}\,\theta(z) = \sum_1^m c_i\, z_i^2 \qquad \text{for all } z = (z_i) \in V'$$

Let $U = \psi^{-1}\theta(V')$ and $\varphi = \theta^{-1}\psi$. Then by DS2, (U,φ) is a chart about p with $\varphi(p) = 0$. Moreover, for all $x = (x_i) \in \varphi(U)$,

$$f\varphi^{-1}(x) = f\psi^{-1}\theta(x)$$

$$= \bar{f}\theta(x) + f(p)$$

$$= \sum_1^m c_i\, x_i^2 + f(p)$$

Since $c_i = \pm 1$, we can rearrange the coordinates to obtain

$$f\varphi^{-1}(x) = -\sum_1^\lambda x_i^2 + \sum_{\lambda+1}^m x_i^2 + f(p) \quad \square$$

Definition Let p be a nondegenerate critical point of the function $f : M^m \to \mathbb{R}$. Then the integer λ obtained in Theorem 4 is called the *index* of p. As in Chap. 4, the index is well defined.

Example $\lambda = 0$ always gives a local minimum and $\lambda = m$ a local maximum. If $m = 2$, then $\lambda = 1$ gives a saddle point. Thus in the case of the function $f : T^2 \to \mathbb{R}$ above, $(-3,0,0)$ has index 0, $(\pm 1,0,0)$ have index 1, and $(3,0,0)$ has index 2.

COROLLARY 5. Nondegenerate critical points of a function are isolated. Thus a function with compact domain and only nondegenerate critical points has only finitely many such points.

Proof: *Isolated* here means that each nondegenerate critical point has a neighborhood containing no other such point. Clearly the neighborhood U of Theorem 4 contains no critical point apart from p, so nondegenerate critical points are isolated.

Suppose C is the set of critical points of a function $f : M \to \mathbb{R}$ having only nondegenerate critical points. For each $p \in C$, choose

9. Critical Points Again

an open set U_p containing p such that U_p contains no other critical point of f . Then $\{U_p | p \in C\} \cup \{M - C\}$ is an open cover of M . Openness of M - C follows from Exercise 1 below. By compactness of M , there is a finite subcover which must contain at most finitely many U_p . Hence C is finite. □

Note that if a function with compact domain has degenerate critical points then it may have infinitely many (even nondegenerate) critical points. The function $f : \mathbb{R} \to \mathbb{R}$ defined by $f(x) = x^6 \sin(1/x)$ has a degenerate critical point at 0 and infinitely many nondegenerate critical points in any neighborhood of 0 . It may be modified to give a function $S^1 \to \mathbb{R}$ with infinitely many nondegenerate critical points.

Definition A *Morse function* on a manifold is a differentiable function all of whose critical points are nondegenerate.

THEOREM 6. Every compact manifold possesses a Morse function.

Proof: Omitted. □

EXERCISES

1. Let $f : M \to N$ be differentiable. Prove that the set

    $\{p \in M \mid f \text{ is regular at } p\}$

    is open in M .

2. Let $f : T^2 \to \mathbb{R}$ be the (Morse!) function considered in the examples above. Find a chart about one of the critical points of f satisfying the requirements of Morse's theorem.

3. Construct a C^2 function $S^1 \to \mathbb{R}$ having infinitely many non-degenerate critical points. [Hints: Modify the example $\mathbb{R} \to \mathbb{R}$ above making use of Lemma 4.1. The graph of a function $X \to Y$ is a subset of $X \times Y$. Stereographic projection enables us to transfer the graph $\{(x, x^6 \sin(1/x)) | x \in \mathbb{R}\}$ as in Fig. 38a to $S^1 \times \mathbb{R}$ as in Fig. 38b. We then need to modify this to Fig. 38c using Lemma 4.1 to give us the required function.]

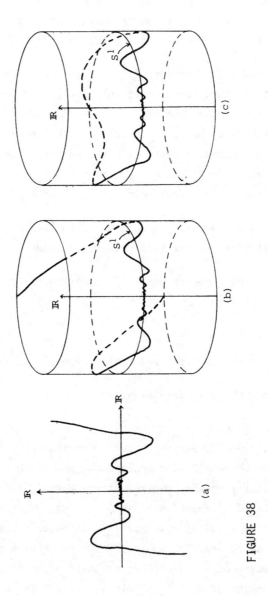

FIGURE 38

9. Critical Points Again

4. Let $f : M \to \mathbb{R}$ be differentiable and let C denote the set of critical points of f. Prove or disprove the statement: if $p \in C$ is degenerate then $p \in \overline{C - \{p\}}$.

5. Let M be a compact manifold. Prove that there is a Morse function $f : M \to \mathbb{R}$ with the property that no two critical points of f have the same value. (Hint: Modify an arbitrary Morse function a small amount in the vicinity of each critical point using Prototype Lemma 4.6. Corollary 5 above ensures that only finitely many modifications are necessary.)

10
VECTOR FIELDS AND INTEGRAL CURVES

Theorem 9.6 tells us that every compact manifold possesses a Morse function. We are going to use Morse functions to study compact manifolds.

Definition Let $f : M \to \mathbb{R}$ be a Morse function. For each $c \in \mathbb{R}$, let $M_c = f^{-1}(c)$. Then M_c is called a *level* of f. The level M_c is called *critical* or *regular* according as c is or is not a critical value of f.

Our program for the classification of orientable surfaces is as follows. Let $f : M^m \to \mathbb{R}$ be a Morse function, where M is compact. Then for some $a,b \in \mathbb{R}$, $f(M) \subset [a,b]$. We study the change in the levels M_c as c varies from a to b, not only the individual levels but also the way in which they fit together. Theorem 9.2 tells us that the regular levels are all $(m-1)$-submanifolds. We can say much more than this. If $[c,d]$ contains no critical values of f, then M_c is diffeomorphic to M_d by a natural diffeomorphism: we can slide M_c through levels of f to M_d. The sliding is along curves called integral curves, which have the property that they are (differentiably!) at right angles to the levels. Integral curves are obtained from vector fields, which are to be studied presently. On the other hand, critical levels need not be manifolds, and a regular level just below a critical level need not be diffeomorphic to a regular level just above the critical level. Thus as we cross a critical level the quality of the regular levels may change. Again integral curves will be used to study this change. Finally, the integral curves also allow us to fit the various levels together.

During our study of the above processes, the reader should keep in mind the Morse function $f : T^2 \to \mathbb{R}$ considered in Chap. 9, since this exhibits all the possible ways in which the levels of a well-chosen Morse function, with domain an orientable surface, change. The regular level ϕ becomes a circle, which splits into two circles. These circles recombine to give a single circle again, which in turn vanishes.

Definition Let M^m be a manifold and let TM denote the disjoint union of the tangent spaces TM_p (as p ranges through M). TM can be topologized in a natural way so that it becomes a 2m-manifold, called the *tangent manifold*, although we will not need the topology here. Define $\pi : TM \to M$ by $\pi(TM_p) = p$, for all $p \in M$.

The requirement that the union be disjoint poses a slight problem above, since $TM_p \cap TM_q \neq \phi$ even if $p \neq q$ [each contains the zero function $C^\infty(M, \mathbb{R}) \to \mathbb{R}$]. There are several ways to overcome this problem, but we prefer to ignore the problem here (see Exercise 1, however). The function π picks out the point at which the tangent vector is based.

Definitions A *vector field* on a manifold M is a function $\xi : M \to TM$ satisfying:

VF1 $\pi\xi = 1$.
VF2 For all $f \in C^\infty(M, \mathbb{R})$, the function $f_\xi : M \to \mathbb{R}$ defined by $f_\xi(p) = \xi(p)(f)$ is C^∞.

Condition VF1 tells us that ξ assigns to each $p \in M$ a tangent vector $\xi(p)$ at p, and condition VF2 is a smoothness condition, in particular a continuity condition, so that if p and q are nearby then so are the vectors $\xi(p)$ and $\xi(q)$. An alternative way of formulating VF2 would be to demand that ξ be a differentiable function when TM is given the differential structure defined in Exercise 1.

10. Vector Fields and Integral Curves

Let $f : M^m \to \mathbb{R}$ be a Morse function. A vector field ξ on M is *gradientlike* for f if and only if

Grad 1. $\xi(p)(f) > 0$ for each regular point p of f.

Grad 2. For every critical point p of f, there is a chart (U, φ) about p so that

$$f\varphi^{-1}(x) = f(p) - \sum_1^\lambda x_i^2 + \sum_{\lambda+1}^m x_i^2 \quad \text{for every } x = (x_i) \in \varphi(U)$$

and the components of $\xi(\varphi^{-1}(x))$ with respect to (U, φ) are

$$(-x_1, \ldots, -x_\lambda, x_{\lambda+1}, \ldots, x_m)$$

Condition Grad 1 ensures that if p is a regular point then $\xi(p)$ is "pointing" in a direction of increasing f (there are many such directions), and Grad 2 tells us that ξ is particularly nice about a critical point. In particular, if p is a critical point, then $\xi(p) = 0$. Note that Grad 1 and Grad 2 are compatible, for if (U, φ) is a chart about the critical point p as in Grad 2, and $q \in U - \{p\}$, say $\varphi(q) = x$, then according to the definition of components given in Chap. 8, we have

$$\xi(q)(f) = \sum_1^\lambda (-x_i) \left.\frac{\partial f}{\partial \varphi_i}\right|_q + \sum_{\lambda+1}^m x_i \left.\frac{\partial f}{\partial \varphi_i}\right|_q$$

$$= -\sum_1^\lambda x_i (-2x_i) + \sum_{\lambda+1}^m x_i \, 2x_i$$

$$= 2 \sum_1^m x_i^2$$

$$> 0$$

Often one discusses a gradientlike vector field without mention of a Morse function. In such a case it is assumed that there is an associated Morse function.

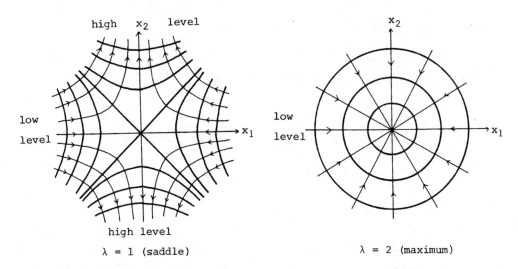

FIGURE 39

Figure 39 illustrates Grad 2 for the cases where $m = 2$ and $\lambda = 1$ or 2. The tangent lines to the thin curves give the directions of the tangent vectors $\xi(q)$ and the thick lines represent levels of f, i.e., sets of the form

$$\varphi^{-1}\{(x_1,\ldots,x_m) \mid -\sum_1^\lambda x_i + \sum_{\lambda+1}^m x_i^2 = \text{constant}\}$$

Examples The simplest example of a vector field is the trivial one: $\xi(p)$ is the zero tangent vector at p. This is not gradientlike for any Morse function.

Define $\xi : \mathbb{R}^m \to T\mathbb{R}^m$ thus: For all $p \in \mathbb{R}^m$, let $\xi(p)$ have components $(1,0,\ldots,0)$ with respect to the chart $(\mathbb{R}^m, 1)$. Then ξ is a vector field on $\mathbb{R}^m$. It is a gradientlike vector field for any smooth function $f : \mathbb{R}^m \to \mathbb{R}$ satisfying $\partial f/\partial x_1 > 0$ everywhere (such a function is a Morse function since it has no critical points).

Define $\xi, \eta : T^2 \to TT^2$ thus. Recall the immersion $\chi : \mathbb{R}^2 - \{0\} \to T^2$ which wraps $\mathbb{R}^2 - \{0\}$ around T^2 and which was

10. Vector Fields and Integral Curves

constructed in Chap. 9. For any $(a,b) \in \mathbb{R}^2 - \{0\}$, let $\xi(\chi(a,b))$ and $\eta(\chi(a,b))$ have components

$$\left(-\frac{b}{\sqrt{a^2+b^2}}, \frac{a}{\sqrt{a^2+b^2}} \right) \qquad \left(\frac{a}{\sqrt{a^2+b^2}}, \frac{b}{\sqrt{a^2+b^2}} \right)$$

respectively, with respect to the charts (U,φ) and (V,ψ) where appropriate. The charts (U,φ) and (V,ψ) here are those constructed in Chap. 5 and considered again in Chap. 9. We verify that ξ satisfies VF2. Suppose $f \in C^\infty(T^2, \mathbb{R})$. Then

$$f_\xi \chi(a,b) = -\frac{b}{\sqrt{a^2+b^2}} \frac{\partial(f\chi)}{\partial a}\bigg|_{(a,b)} + \frac{a}{\sqrt{a^2+b^2}} \frac{\partial(f\chi)}{\partial b}\bigg|_{(a,b)}$$

Since f is C^∞, so is $f\chi$, and hence $f_\xi \chi$ and f_ξ are C^∞. Neither of these vector fields is gradientlike for a Morse function, since they are both nonzero throughout T^2 but every Morse function on T^2 has critical points. The vector field ξ determines a unit longitudinal flow around the torus, i.e., $\xi(p)$ is the velocity vector of a unit speed curve along the longitudinal line through p, and η determines a unit latitudinal flow.

Recall the Morse function $f : T^2 \to \mathbb{R}$ given by $f(x,y,z) = x$, considered in Chap. 9. For each $p \in T^2$, let $\xi(p)$ be the projection onto the tangent plane to T^2 at p of a vector of appropriate length in the positive x direction. If the length is well chosen, then ξ is a gradientlike vector field for f. In Fig. 40, $\xi(p)$ is the velocity vector to the flow curve at p. The length of $\xi(p)$ is much greater than the length of $\xi(q)$. The flow curves are the integral curves of Theorem 2.

The gradientlike vector field on T^2 just considered is a good example to keep in mind when trying to understand the relationship of a gradientlike vector field and its Morse function. Think of water flowing on any landscape. Taking height as the Morse function, minus the velocity of flow of the water determines a gradientlike

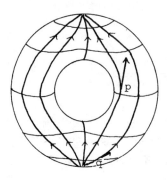

FIGURE 40

vector field for the Morse function. The steeper the landscape, the faster the flow, hence the longer the vector at the point in question. Around a hilltop, saddle, or basin bottom, the flow will be very slow, 0 at the point, corresponding to Grad 2. At a critical point the water "does not know" which direction to flow, corresponding to $\xi(p) = 0$.

THEOREM 1. Every Morse function with compact domain has a gradient-like vector field.

Proof: Let $f : M^m \longrightarrow \mathbb{R}$ be a Morse function, M compact. We firstly outline the proof. Grad 2 dictates to us how the required vector field ξ must be defined about each of the critical points. Morse's theorem (9.4) gives us a particularly nice chart about each critical point. We use this chart to give us a vector field in a neighborhood of each critical point (much as in Fig. 39). As in the proof of Theorem 7.2, these vector fields are extended over the whole of M, being 0 except in a neighborhood of the critical point. In this way, we obtain several vector fields, one for each critical point, so that a particular one satisfies Grad 2 at its own critical point, and is zero away from its critical point. Adding these together gives us a vector field satisfying Grad 2 but perhaps not Grad 1, since a vector may still be 0 at regular points which are far from critical points.

10. Vector Fields and Integral Curves

Now we appeal to Theorem 9.1 to tell us a direction in which f is increasing at regular points. In this way we obtain some more vector fields which are 0 in a neighborhood of each critical point, and elsewhere are either 0 or point in a direction of increasing f. Using compactness, we find we need only finitely many such vector fields to ensure that at each regular point at least one of the vector fields points in a direction of increasing f. Add all of these vector fields together to get ξ. Now for the details.

Suppose the critical points of f are $p_1, \ldots, p_k$, with respective indices $\lambda_1, \ldots, \lambda_k$. For each $j = 1, \ldots, k$, let (U_j, φ_j) be a chart about p_j given by Morse's theorem, and let U_j' be another open neighborhood of p_j such that $\mathrm{Cl}\, U_j' \subset U_j$. The sets U_j' form a kind of protective barrier around the points p_j. They ensure that when we form the sum ξ of the various constructed vector fields ξ_i, the only one which is nonzero in some neighborhood of p_j is ξ_j. By ensuring that ξ_j satisfies Grad 2 near p_j, we ensure that ξ does as well.

For each $p \in M - \bigcup_{j=1}^{k} U_j$, use Theorem 9.1 to find a chart (U_p, φ_p) about p satisfying:

i. For all $j = 1, \ldots, k$, $U_j' \cap U_p = \emptyset$.
ii. For all $x = (x_i) \in \varphi_p(U_p)$, $f\varphi_p^{-1}(x) = x_1 +$ constant.

Now $\bigcup_{j=1}^{k} U_j$ is open, so $M - \bigcup U_j$ is closed, hence, by Theorem 3.5, is compact. Thus the open cover $\{U_p \mid p \in M - \bigcup_{j=1}^{k} U_j\}$ has a finite subcover which we rename $\{U_j \mid j = k+1, \ldots, \ell\}$. Thus we have an open cover $\{U_j \mid j = 1, \ldots, \ell\}$ of M so that, in addition to the Morse's theorem property of U_j for $j \leq k$, we have

i. For all $i = 1, \ldots, k$, and all $j = k+1, \ldots, \ell$, $U_i' \cap U_j = \emptyset$.
ii. For all $j = k+1, \ldots, \ell$, and all $x = (x_i) \in \varphi_j(U_j)$, $f\varphi_j^{-1}(x) = x_1 +$ constant.

There is another open cover $\{V_j \mid j = 1, \ldots, \ell\}$ of M so that for all j, V_j is open and $\mathrm{Cl}\, V_j \subset U_j$ (why?).

Using Lemma 4.1, we may find smooth functions $\pi_j : M \to \mathbb{R}$ so that $\pi_j(V_j) = 1$ and $\text{Cl } \pi_j^{-1}((0,1]) \subset U_j$, for all $j = 1, \ldots, \ell$ (cf. the proof of Theorem 7.2).

We now define vector fields ξ_j, $j = 1, \ldots, \ell$, on M such that for $j \leq k$, ξ_j is of the form required of the sought after vector field on the neighborhood V_j of p_j, and for $j > k$, ξ_j satisfies Grad 1 within V_j. The required vector field will be $\xi = \sum_{j=1}^{\ell} \xi_j$. Care must be taken to ensure that the ξ_i ($i \neq j$) do not interfere too much with ξ_j near p_j. This is where the π_j are used to taper ξ_i off to 0 outside U_i.

Suppose $p \in M$. We must define $\xi_j(p)$. If $\pi_j(p) = 0$, just set $\xi_j(p) = 0$. If $p \in U_j$, say $\varphi_j(p) = x = (x_i)$, we consider two cases separately. If $j \leq k$, let $\xi_j(p)$ have components

$$(-x_1 \pi_j(p), \ldots, -x_{\lambda_j} \pi_j(p), x_{\lambda_j+1} \pi_j(p), \ldots, x_m \pi_j(p))$$

with respect to the chart (U_j, φ_j); if $j > k$, let $\xi_j(p)$ have components $(\pi_j(p), 0, \ldots, 0)$ with respect to the chart (U_j, φ_j).

That each ξ_j satisfies VF2 follows in much the same way as the smoothness of the functions $\hat{\varphi}_i$ in the proof of Theorem 7.2. Thus each ξ_j, and hence ξ, is a vector field.

In the neighborhood $U_j' \cap V_j$ of p_j, ξ reduces to ξ_j, which satisfies Grad 2 in this neighborhood. Hence ξ satisfies Grad 2.

It remains to verify Grad 1. Suppose $p \in U_j$, say $\varphi_j(p) = (x_1, \ldots, x_m)$. Then

$$\xi_j(p)(f) = \begin{cases} 2\pi_j(p) \sum_{i=1}^{m} x_i^2 & \text{if } j \leq k \\ \\ \pi_j(p) & \text{if } j > k \end{cases}$$

In particular, if $p \in V_j$ and p is regular, then $\xi_j(p)(f) > 0$. Furthermore, for all $p \in M$ and all $i = 1, \ldots, \ell$, $\xi_i(p)(f) \geq 0$. Thus, since $\{V_i \mid i = 1, \ldots, \ell\}$ is a cover of M, $\xi(p)(f) > 0$ when p is a regular point, so Grad 1 is satisfied. □

10. Vector Fields and Integral Curves

In order to apply vector fields to the study of the quality of a manifold between adjacent critical levels, we must extend the notion of a manifold to include the possibility of a boundary.

Definition An *m-manifold with boundary* is a Hausdorff space M^m each point of which has an open neighborhood homeomorphic either to $\mathbb{R}^m$ or to the *half space*

$$H^m = \{(x_1,\ldots,x_m) \in \mathbb{R}^m \mid x_m \geq 0\}$$

If U is an open subset of M and $\varphi : U \to \mathbb{R}^m$ is an embedding, then, just as in the case of a manifold, (U,φ) is called a (coordinate) chart. There is an important subset of a manifold with boundary. Let

$$\partial M = \{x \in M \mid x \text{ has no neighborhood homeomorphic to } \mathbb{R}^m\}$$

Then ∂M is called the *boundary* of M. Note that ∂M is an $(m-1)$-manifold. If $\partial M = \emptyset$, then M is a manifold.

Given a manifold with boundary, say M, we can define a differential structure of class C^r on M exactly as in Chap. 5, and again as in Chap. 5, we can easily extend the notion of differentiability of functions.

If M is a manifold with boundary, then a *Morse function* $f : M \to \mathbb{R}$ is a smooth function whose critical points are nondegenerate and are contained in $M - \partial M$.

We will also need the notion of the product of a manifold $(M^m, \mathcal{D})$ and a manifold with boundary (N^n, E). $(M \times N, \mathcal{D} \times E)$ is an $(m+n)$-manifold with boundary, where

$$M \times N = \{(x,y) \mid x \in M, y \in N\}$$

and $\mathcal{D} \times E$ is the differential structure with basis

$$\{(U \times V, \varphi \times \psi) \mid (U,\varphi) \in \mathcal{D}, (V,\psi) \in E\}$$

$$(\varphi \times \psi)(u,v) = (\varphi(u), \psi(v)) \in \mathbb{R}^m \times \mathbb{R}^n = \mathbb{R}^{m+n}$$

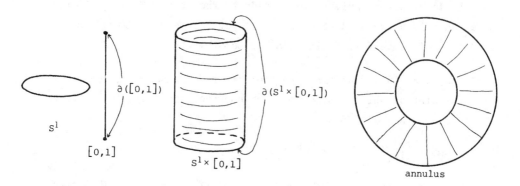

FIGURE 41

Notice that $\partial(M \times N) = M \times \partial N$. $N \times M$ is defined in the obvious way. If M and N are both manifolds with boundary, then there is difficulty in imposing a differential structure near $\partial M \times \partial N$, but fortunately we will not encounter this situation.

Examples Manifolds are manifolds with boundary—the boundary being empty (!). H^m is a manifold with boundary; $\partial H^m = \mathbb{R}^{m-1}$. B^m is a manifold with boundary; $\partial B^m = S^{m-1}$. H^m and B^m both have natural differential structures.

If S^1 and $[0,1]$ have their usual structures, then $S^1 \times [0,1]$ is a cylinder, which is diffeomorphic to the annulus $\{x \in \mathbb{R}^2 \mid 1 \leq |x| \leq 2\}$ (see Fig. 41). $S^1 \times S^1$ is diffeomorphic to T^2.

$[0,1]$ has a natural differential structure, but how can the corners of the square $[0,1] \times [0,1]$ be smoothed out? Fig. 42 illustrates the problem. Topologically this can be done; i.e., if M and N are both topological manifolds with boundary then so is $M \times N$, with $\partial(M \times N) = (\partial M \times N) \cup (M \times \partial N)$.

THEOREM 2. Let $f : M^m \to \mathbb{R}$ be a Morse function, M a compact manifold, and suppose that $[c,d]$ contains no critical values of f. Then $f^{-1}([c,d])$ is diffeomorphic to $M_c \times [c,d]$.

10. *Vector Fields and Integral Curves* 133

[0,1] × [0,1]:
Sharp at the
4 corners.

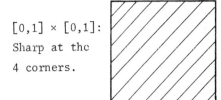

FIGURE 42

Proof: The idea of the proof is as follows. Construction of a diffeomorphism from $M_c \times [c,d]$ to $f^{-1}([c,d])$ involves assigning to each ordered pair $(p,t) \in M_c \times [c,d]$ a point of $f^{-1}([c,d])$. For a fixed $p \in M_c$, $t \mapsto (p,t)$ determines a curve in $M_c \times [c,d]$ as t varies through $[c,d]$. We will construct curves $\gamma_p : [c,d] \to f^{-1}([c,d])$, for each $p \in M_c$, such that $f\gamma_p(t) = t$. These curves fit together smoothly to give us the required diffeomorphism $\Gamma : M_c \times [c,d] \to f^{-1}([c,d])$ defined by $\Gamma(p,t) = \gamma_p(t)$. The curves themselves will be obtained by integrating a gradientlike vector field for f ; hence they are called integral curves, and will possess the property that the velocity vector of a curve at a point is the tangent vector given by the vector field. The reader should try to draw a few such curves on the torus, where f is the function of our standard example and $c = -\frac{1}{2}$, $d = \frac{1}{2}$, say.

Let ξ be a gradientlike vector field for f , the existence of which is assured by Theorem 1.

Since for all $p \in f^{-1}([c,d])$, $\xi(p)(f) > 0$, we may assume that $\xi(p)(f) = 1$ for all such p . To attain this situation, we may multiply $\xi(p)$ by $1/\xi(p)(f)$ for $p \in f^{-1}([c,d])$, using Lemma 4.1 to taper off this alteration of ξ outside $f^{-1}([c,d])$. [This last point is not really necessary as we do not need ξ outside an arbitrary neighborhood of $f^{-1}([c,d])$.] Because of VF2, $\xi(p)(f)$ depends smoothly upon p , so the product $[1/\xi(p)(f)]\, \xi(p)$ will still depend smoothly upon p , i.e., will satisfy VF2 .

Suppose (U,φ) is any chart in the structure of M. If $x \in \varphi(U)$ and $F(x)$ denotes the $1 \times m$ vector of components of $\xi(\varphi^{-1}(x))$ with respect to (U,φ), then we obtain a system of differential equations

$$x' = F(x)$$

where x' denotes differentiation with respect to some parameter t. The fundamental existence and uniqueness theorem for ordinary differential equations gives us a family of curves in $\varphi(U)$ such that the tangent to any curve at any point x is just the vector $F(x)$. Transferring this family to M via φ gives us a family of curves in U such that for all $p \in U$, the velocity vector of the curve passing through p is $\xi(p)$.

Let δ_p denote the curve in U through p. Then

$$(f\delta_p)'(t) = \xi(\delta_p(t))(f)$$

On the left-hand side of this equation, we have the derivative of $f\delta_p$, which is a function whose domain and range both lie in $\mathbb{R}$. Thus $(f\delta_p)'(t)$ is just ordinary one variable differentiation. If $\delta_p(t) \in f^{-1}([c,d])$, then $\xi(\delta_p(t))(f) = 1$, so that

$$(f\delta_p)'(t) = 1$$

and hence within $f^{-1}([c,d])$, we have

$$f\delta_p(t) = t + \text{constant}$$

By a change of parameter we can eliminate the constant to obtain curves γ_p such that $f\gamma_p(t) = t$ whenever $\gamma_p(t)$ lies within $f^{-1}([c,d])$, still retaining the property that the velocity vector of γ_p at any point is the tangent vector given by ξ at that point.

The uniqueness part of the fundamental theorem tells us that the curves γ_p are independent of the choice of chart (U,φ) except possibly for the domain of γ_p. Such curves are called *integral curves* for ξ.

10. Vector Fields and Integral Curves

Consider the integral curves within $f^{-1}([c,d])$. The domains of these curves all lie in $[c,d]$. Each such curve can be extended over a maximal subinterval of $[c,d]$, which we claim to be $[c,d]$ itself. Indeed, suppose γ is an integral curve extended over a maximal subinterval I of $[c,d]$. Suppose that b is the upper endpoint of I. In two steps we show that $b \in I$.

Step I. There exists $p \in M_b$ such that for every neighborhood N of p, $N \cap \gamma(I) \neq \emptyset$. In fact, if for each $p \in M_b$, there is a neighborhood N_p of p with $N \cap \gamma(I) = \emptyset$, then $N = \cup \{N_p \mid p \in M_b\}$ is a neighborhood of M_b which does not meet $\gamma(I)$. By Exercise 6, any neighborhood of M_b contains M_{b-r} for all sufficiently small positive r. Since $f\gamma(t) = t$, we have $\gamma(b - r) \in M_{b-r}$, so that if we have taken r small enough so that $M_{b-r} \subset N$, we have found a point of $N \cap \gamma(I)$, a contradiction. Thus there is such a $p \in M_b$ as claimed.

Step II. Let $p \in M_b$ be as in Step I. By VF2 the integral curve through any point near p lies near γ_p. In particular, since γ_p extends to a level higher than M_b, any integral curve near p must also extend to a level higher than M_b. In particular, γ is near p, so γ must reach the level M_b, i.e., $b \in I$.

The last part of Step II actually showed us more, viz., that γ extends beyond b. Thus $b = d$. Similarly, the lower endpoint of I is c, and $c \in I$; thus the domain of γ is $[c,d]$.

Thus for every $p \in M_c$, we have found a curve $\gamma_p : [c,d] \to M$ satisfying the following:

i. The velocity vector of γ_p at any point is the tangent vector given by ξ at that point.
ii. For every $t \in [c,d]$, $f\gamma_p(t) = t$.

Now define the required diffeomorphism

$$\Gamma : M_c \times [c,d] \to f^{-1}([c,d])$$

by $\Gamma(p,t) = \gamma_p(t)$.

Differentiability of Γ follows from VF2 via condition ii above. □

EXERCISES

1. Let $(M^m, \mathcal{D})$ be a C^r manifold, let

   $$TM = \{(p,v) \mid p \in M, v \in TM_p\}$$

   and define $\pi : TM \to M$ by $\pi(p,v) = p$. For every $(U, \varphi) \in \mathcal{D}$, let $\hat{U} = \pi^{-1}(U)$, and define $\hat{\varphi} : \hat{U} \to \mathbb{R}^{2m}$ by

   $$\hat{\varphi}(p,v) = (\varphi(p), (\alpha_1, \ldots, \alpha_m)) \quad \text{for every } (p,v) \in \hat{U}$$

   where $(\alpha_1, \ldots, \alpha_m)$ are the components of v with respect to (U, φ). Prove that $\{(\hat{U}, \hat{\varphi}) \mid (U, \varphi) \in \mathcal{D}\}$ is a basis for a differential structure of class C^{r-1} on TM.

2. Let f, ξ, p, and (U, φ) be as in Grad 2. Rename x_j ($j \geq \lambda + 1$) as $y_{j-\lambda}$ and set $x = (x_1, \ldots, x_\lambda)$, $y = (y_1, \ldots, y_{m-\lambda})$; thus $\xi(\varphi^{-1}(x,y))$ has components $(-x_1, \ldots, -x_\lambda, y_1, \ldots, y_{m-\lambda})$. Determine the integral curves for ξ within (U, φ), i.e., solve the differential equations $(dx/dt, dy/dt) = (-x, y)$. Show that (unless $\lambda = 0$ or m) the integral curves satisfy the equation $|x||y| = $ constant; thus justify Fig. 39.

3. The function $f : S^2 \to \mathbb{R}$ defined by $f(x,y,z) = z$ is a Morse function (why?). Construct gradientlike vector fields for f whose integral curves

   (a) Are the lines of longitude

   (b) Spiral around S^2 finitely many times

4. Prove that $S^1 \times [0,1]$ and $\{x \in \mathbb{R}^2 \mid 1 \leq |x| \leq 2\}$ are diffeomorphic.

5. Why is Γ^{-1} in the proof of Theorem 2 differentiable?
   (Hint: Appeal to the inverse function theorem.)

6. Let $f : M \to \mathbb{R}$ be a Morse function with M compact. Let $b \in \mathbb{R}$, and let U be a neighborhood of M_b in M. Prove that there exists $r > 0$ such that $f^{-1}([b-r, b+r]) \subset U$.
   (Hint: Suppose not. Then there is a sequence (x_n) of points of $M - U$ so that for all n, $|f(x_n) - b| < 1/n$. If U is open then $M - U$ is compact and hence (x_n) has a convergent subsequence

10. *Vector Fields and Integral Curves* 137

whose limit also lies in $M - U$. Since $f(x_n) \to b$, the limit has value b under f, leading to a contradiction.)

7. Let M^m be a compact manifold which admits a Morse function having exactly two critical points. Prove that M is homeomorphic to S^m.

11
SURGERY

Theorem 2 of Chap. 10 tells us that the structure of a manifold does not change between two levels when the portion of the manifold between the levels contains only regular points. As we cross a critical level, however, the quality of the level might change. Consider, for example, the Morse function $f : T^2 \longrightarrow \mathbb{R}$ given by $f(x,y,z) = x$, studied in Chap. 9. This function has four critical levels, $f^{-1}(\pm 1)$ and $f^{-1}(\pm 3)$. Levels below $f^{-1}(-3)$ are $\emptyset$; levels between $f^{-1}(-3)$ and $f^{-1}(-1)$ are (diffeomorphic to) S^1; levels between $f^{-1}(-1)$ and $f^{-1}(1)$ consist of two circles; levels between $f^{-1}(1)$ and $f^{-1}(3)$ are S^1 again; levels above $f^{-1}(3)$ are $\emptyset$ again. Thus the topological type of the level may change as we pass through a critical level (although not necessarily; cf. the twisting surgery of Chap. 13). It turns out, as we will discover in Chap. 13, that the above four changes are essentially the only ways that the levels of an orientable 2-manifold can change as we cross a critical level. Thus although we will be working in a much more general context in the next two chapters, the above situation should be strongly borne in mind, especially in Chap. 12.

Whereas in Chap. 10 we were able to determine the structure of a manifold between two regular levels with no intervening critical level quite quickly, it is somewhat more difficult when there is an intervening critical level. At least, using Exercise 9.5, we are able to assume that the intervening critical level contains only one critical point. In this chapter we will develop a technique which will enable us to determine how a supercritical level is related to a subcritical level. That it does enable us to do this will not become evident until Chap. 12, where we actually determine the whole of the manifold between the two levels. Thus in this chapter we will not have any use for Morse functions. However, the reader might like to anticipate how the technique developed here relates to Morse functions, especially to the standard example.

A change in levels as we pass over a critical level involves a process known as surgery. We begin with a description of surgery which ignores differentiability. In its simplest form, surgery exploits the following fact, illustrated by Fig. 43:

$$\partial(S^{m-1} \times B^n) = S^{m-1} \times S^{n-1} = \partial(B^m \times S^{n-1})$$

Thus if we have a manifold with boundary, whose boundary is (diffeomorphic to) $S^{m-1} \times S^{n-1}$ then we can extend the manifold by sewing either $S^{m-1} \times B^n$ or $B^m \times S^{n-1}$ along the boundary. Surgery is, essentially, the process which goes from one of these manifolds to the other. More precisely, suppose that M^{m+n-1} is a manifold (no boundary) and $e : S^{m-1} \times B^n \to M$ is an embedding. Topologically, surgery removes $e(S^{m-1} \times \text{Int } B^n)$ by cutting along $e(S^{m-1} \times S^{n-1})$, and replaces it by $\text{Int } B^m \times S^{n-1}$, the attachment being along $e(S^{m-1} \times S^{n-1})$.

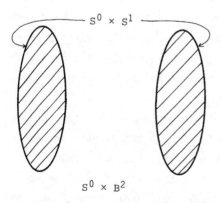
$S^0 \times B^2$

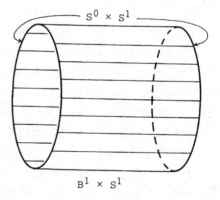
$B^1 \times S^1$

FIGURE 43

11. Surgery

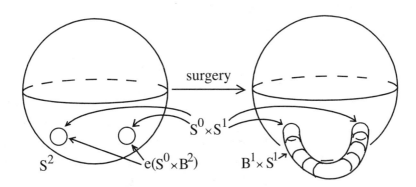

FIGURE 44

Figure 44 illustrates a possible surgery when $m = 1$, $n = 2$, and $M^2 = S^2$. The effect of this surgery on S^2 is to change it into T^2 (up to homeomorphism!). Note that, in general, we could perform surgery along any manifold which is simultaneously the boundary of two manifolds with boundary.

As suggested above, the main technical difficulty is the desire to make the new manifold smooth at $e(S^{m-1} \times S^{n-1})$. The problem is that any chart about a point on $e(S^{m-1} \times S^{n-1})$ in the new manifold must be made up of parts of two charts, one from M and one from $B^m \times S^{n-1}$. This is overcome by using the fact that differentiability is a local phenomenon, so we make the overlap of the two pieces which comprise the new manifold an open set, with the two differential structures on the overlap being compatible (in the sense of DS1 of Chap. 5).

We begin by making the process of attaching two manifolds together more precise.

Definition Let M^m and N^m be two differentiable manifolds with $M \cap N = \emptyset$. Let $h : O \twoheadrightarrow N$ be an embedding of an open subset of M in N satisfying the condition that for all $x \in M$ and all $y \in N - \{h(x)\}$, there is a neighborhood U of x in M and a neighborhood V of y in N such that $h(O \cap U) \cap V = \emptyset$.

The *adjunction manifold* obtained from M and N via h is the manifold $(M \cup_h N, \mathcal{D})$ defined as follows:

$$M \cup_h N = (M - O) \cup [N - h(O)] \cup \{(x, h(x)) \mid x \in O\}$$

Thus a topological manifold, $M \cup_h N$ consists of three pieces, and when you construct an adjunction manifold later you should identify each of these pieces and see how they fit together. The last piece is the graph of the embedding h. It has the same dimension as M and N, viz., m. In fact, this piece is diffeomorphic to O by the diffeomorphism which takes $x \in O$ to the point $(x, h(x))$ in the graph. Thus we may think of this piece as being available to restore O to the first piece. On the other hand, the last piece is also diffeomorphic to $h(O)$ by the diffeomorphism which takes the point $y \in h(O)$ to the point $(h^{-1}(y), y)$ in the graph. Thus we may also think of this piece as being available to restore $h(O)$ to the second piece. In this way the two manifolds are sewn together. Using the above, we are able to think of M and N as being submanifolds of $M \cup_h N$, and this is how we construct the differential structure (and, indeed, the topology) on $M \cup_h N$. Define $i : M \to M \cup_h N$ and $j : N \to M \cup_h N$, natural inclusions of M and N as submanifolds of $M \cup_h N$, by

$$i(x) = \begin{cases} x & \text{if } x \in M - O \\ (x, h(x)) & \text{if } x \in O \end{cases}$$

$$j(y) = \begin{cases} y & \text{if } y \in N - h(O) \\ (h^{-1}(y), y) & \text{if } y \in h(O) \end{cases}$$

Suppose A is a basis for the differential structure of M and B is a basis for the differential structure of N. Let

$$C = \{(i(U), \varphi i^{-1}) \mid (U, \varphi) \in A\} \cup \{(j(V), \psi j^{-1}) \mid (V, \psi) \in B\}$$

According to the Lemma 1 below, C is a basis for a differential structure $\mathcal{D}$ on $M \cup_h N$.

11. Surgery

The effect of the adjunction is to sew the manifolds M and N together by identifying each $x \in O$ with $h(x) \in h(O)$. The maps i and j in effect carry the differential structures of M and N, respectively, onto $M \cup_h N$. Compatibility of the differential structure follows from the fact that h is differentiable.

LEMMA 1. *The collection C of charts above is a basis for a differential structure on $M \cup_h N$.*

Proof: We must verify two facts. The open sets in C cover $M \cup_h N$, and the charts in C satisfy DS1 of Chap. 5.

Since A is a basis for a differential structure on M, $\{i(U) \mid (U,\varphi) \in A\}$ covers $i(M)$; similarly, $\{j(V) \mid (V,\psi) \in B\}$ covers $j(N)$. Since $i(M) \cup j(N) = M \cup_h N$, we see that $M \cup_h N$ is covered by the open sets in C.

Any two charts from $\{(i(U), \varphi i^{-1}) \mid (U,\varphi) \in A\}$ or from $\{(j(V), \psi j^{-1}) \mid (V,\psi) \in B\}$ clearly satisfy the compatibility condition DS1. Suppose that $(U,\varphi) \in A$ and $(V,\psi) \in B$. To verify that $(\psi j^{-1})(\varphi i^{-1})^{-1}$ is differentiable on $\varphi i^{-1}(i(U) \cap j(V))$, note that on O we have $j^{-1}i = h$, since for $x \in O$,

$$j^{-1}i(x) = j^{-1}(x, h(x)) = h(x)$$

Thus we have

$$(\psi j^{-1})(\varphi i^{-1})^{-1} = \psi j^{-1} i \varphi^{-1} = \psi h \varphi^{-1}$$

which is differentiable since h is an embedding. Similarly we have

$$(\varphi i^{-1})(\psi j^{-1})^{-1} = \varphi i^{-1} j \psi^{-1} = \varphi h^{-1} \psi^{-1}$$

which is differentiable for the same reason. □

Strictly we should verify that $M \cup_h N$ is a topological manifold before discussing any possible differential structure thereon. Thus we should impose a topology on $M \cup_h N$ and verify that it satisfies the conditions for a topological manifold. The topology is determined by applying Exercise 3.9 to the basis C for the differential structure. That $M \cup_h N$ is Hausdorff follows from the condition we imposed on our embedding h.

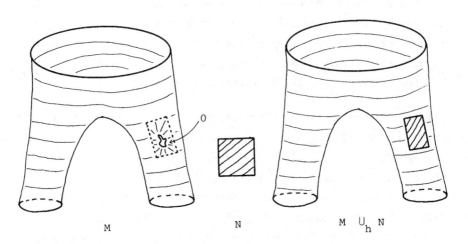

FIGURE 45

Examples Let M^m be any manifold, $M \cap \mathbb{R}^m = \emptyset$, let O be an open subset of M, and $h : O \to \mathbb{R}^m$ a diffeomorphism. Then $M \cup_h \mathbb{R}^m$ is essentially just M.

Let M be an old pair of jeans with a hole in one knee, and let N be a rectangular patch. Let O be the overlap of the jeans with the patch and let $h : O \to N$ assign to each $x \in O$ the point $h(x)$ to which x is stitched in effecting a repair. Then $M \cup_h N$ is the repaired jeans; see Fig. 45. Using this analogy, we will often refer to N as a patch. The analogy is quite apt, as in repairing the hole we usually require the overlap to consist of more than the rim of the hole.

Let $M^1 = \{(x,y) \in \mathbb{R}^2 \mid (x-3)^2 + y^2 = 1\}$; then $M \cap S^1 = \emptyset$. Let $O = \{(x,y) \in M \mid y > 0\}$, and define $h : O \to S^1$ by $h(x,y) = (x-3, y)$. Then h is an embedding but does not satisfy the separation condition. We can still define $M \cup_h S^1$, illustrated in Fig. 46, the result being a differentiable manifold except for the Hausdorff condition. The points $i(4,0)$ and $j(1,0)$ do not have disjoint neighborhoods.

11. Surgery

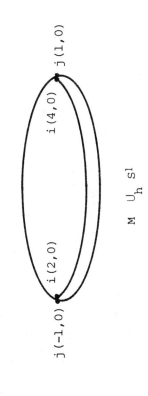

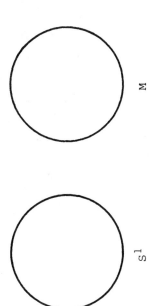

FIGURE 46

We return now to the process of surgery. In our description above we had an embedding $e : S^{m-1} \times B^n \to M^{m+n-1}$ which was used to remove $e(S^{m-1} \times \text{Int } B^n)$ from M. The set $B^m \times S^{n-1}$ was sewn to the result by sewing $e(x,y) \in e(S^{m-1} \times S^{n-1})$ to $(x,y) \in S^{m-1} \times S^{n-1}$. In the context of our adjunction space $M \cup_h N$, we have replaced M by $M - e(S^{m-1} \times \text{Int } B^n)$, h by $e^{-1} \mid e(S^{m-1} \times S^{n-1}) : e(S^{m-1} \times S^{n-1}) \to B^m \times S^{n-1}$, and N by $B^m \times S^{n-1}$. Two requirements of the definition are missing. We have manifolds with boundary rather than without, and the domain of the sewing function is not an open subset of the appropriate manifold. Instead of removing all of $e(S^{m-1} \times \text{Int } B^n)$, we might remove only part of it, say $e(S^{m-1} \times \tfrac{1}{2}B^n)$. Note that $M - e(S^{m-1} \times \text{Int } B^n)$ is much the same as $M - e(S^{m-1} \times \tfrac{1}{2}B^n)$, although the latter does not have a boundary. Having removed only part of $e(S^{m-1} \times \text{Int } B^n)$, we then do not need the boundary of $B^m \times S^{n-1}$ as part of the patch. This removes the first objection. All that remains is to specify a diffeomorphism between the open subset $e(S^{m-1} \times (\text{Int } B^n - \tfrac{1}{2}B^n))$ of $M - e(S^{m-1} \times \tfrac{1}{2}B^n)$ and the open subset $(\text{Int } B^m - \tfrac{1}{2}B^m) \times S^{n-1}$ of $\text{Int } B^m \times S^{n-1}$ which satisfies the condition of the definition. This is done by modifying the embedding which we used to do the sewing. Whereas it was originally $e^{-1} \mid e(S^{m-1} \times S^{n-1})$, we must now extend it over $e(S^{m-1} \times (\text{Int } B^n - \tfrac{1}{2}B^n))$. This is done by following radial lines. More precisely, for $u \in S^{m-1}$ and $v \in S^{n-1}$, the image under e of the radial line $\{(u,rv) \mid r \in (\tfrac{1}{2},1)\}$ is sewn to the radial line $\{(ru,v) \mid r \in (\tfrac{1}{2},1)\}$. In effect we treat $\text{Int } B^n - \tfrac{1}{2}B^n$ as the product $(\tfrac{1}{2},1) \times S^{n-1}$, so that $S^{m-1} \times (\text{Int } B^n - \tfrac{1}{2}B^n)$ is like $S^{m-1} \times (\tfrac{1}{2},1) \times S^{n-1}$. The factor $(\tfrac{1}{2},1)$ is then transferred to the sphere S^{m-1} to give $(\text{Int } B^m - \tfrac{1}{2}B^m) \times S^{n-1}$, since $S^{m-1} \times (\tfrac{1}{2},1)$ is like $\text{Int } B^m - \tfrac{1}{2}B^m$. The diffeomorphism

$$S^{m-1} \times (\text{Int } B^n - \tfrac{1}{2}B^n) \to (\text{Int } B^m - \tfrac{1}{2}B^m) \times S^{n-1}$$

defined by

$$(u,rv) \mapsto (ru,v)$$

describes this exactly. Fig. 47 illustrates the process. There is

11. Surgery

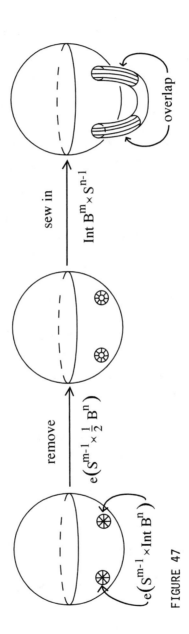

FIGURE 47

a refinement. Instead of stopping at the ball of radius ½, we might cut out only $S^{m-1} \times \{0\}$, sewing in the same patch so that the only new part is $\{0\} \times S^{n-1}$. Thus in $S^{m-1} \times (\text{Int } B^n - \{0\})$, which is like $S^{m-1} \times (0,1) \times S^{n-1}$, we transfer the factor $(0,1)$ to the sphere to obtain $(\text{Int } B^m - \{0\}) \times S^{n-1}$. This is described exactly by the diffeomorphism

$$\alpha : S^{m-1} \times (\text{Int } B^n - \{0\}) \longrightarrow (\text{Int } B^m - \{0\}) \times S^{n-1}$$

defined by $\alpha(u, rv) = (ru, v)$ for $u \in S^{m-1}$, $v \in S^{n-1}$, $r \in (0,1)$. Finally, we no longer need $e(S^{m-1} \times S^{n-1})$, so we assume that e has domain $S^{m-1} \times \text{Int } B^n$.

Definition Let M^{m+n-1} be a manifold and $e : S^{m-1} \times \text{Int } B^n \to M$ be an embedding. Denote by $X(M,e)$ the adjunction manifold

$$[M - e(S^{m-1} \times \{0\})] \cup_{\alpha e^{-1}} (\text{Int } B^m \times S^{n-1})$$

Any manifold diffeomorphic to $X(M,e)$ is said to be obtainable from M by a *surgery of type* (m,n).

Remark The above definition is still valid if either $m = 0$ or $n = 0$. In either case we need to interpret S^{-1}. Recall that S^{n-1} consists of all points of $\mathbb{R}^n$ at distance 1 from the origin, so S^{-1} should contain all points of $\mathbb{R}^0$ at distance 1 from the origin. The natural inclusion $\mathbb{R}^n \subset \mathbb{R}^{n+1}$ for $n > 0$ suggests that $\mathbb{R}^0$ should be $\{0\}$, a single point, in which case $S^{-1} = \emptyset$. Note also that $B^0 = \{0\}$, being the set of points of $\mathbb{R}^0$ at distance at most 1 from the origin.

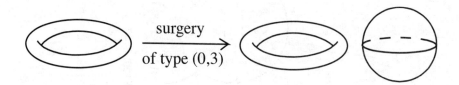

FIGURE 48

11. Surgery

Examples (i) $m = 0$. In this case we require an embedding e of $S^{-1} \times \text{Int } B^n = \emptyset \times \text{Int } B^n = \emptyset$ in M^{n-1}. The surgery then removes nothing from M, sewing in $\text{Int } B^0 \times S^{n-1} \approx S^{n-1}$ by use of the diffeomorphism αe^{-1}. Since the domain of αe^{-1} is empty, the effect is to add a sphere of dimension $n - 1$ to the manifold. No point of the sphere is near M nor is any point of M near the added sphere (see Fig. 48).

(ii) $n = 0$. In this case we require an embedding e of $S^{m-1} \times \text{Int } B^0 \approx S^{m-1}$ in M^{m-1}. Thus such a surgery cannot be performed if M does not contain a submanifold diffeomorphic to S^{m-1}; for example, (3,0) surgery cannot be performed on T^2 since T^2 has no submanifold diffeomorphic to S^2. Surgery of type $(m,0)$ removes a sphere of dimension $m - 1$, replacing it by nothing. Thus Fig. 48 illustrates also a surgery of type (3,0) if we go from right to left.

(iii) $m = n = 1$. In this case we require an embedding e of $S^0 \times \text{Int } B^1$, i.e., of a pair of open intervals, in M^1. The surgery removes the pair of points $e(S^0 \times \{0\})$ and replaces it by another pair of points, $\{0\} \times S^0$. Consider the particular case where M consists of a pair of circles and e embeds an interval in each, as in Fig. 49. The effect of the surgery is to join the two circles into a single circle. Note that this is what happens when we jump from a level just below the critical point $(1,0,0)$ of our standing example $f : T^2 \to \mathbb{R}$ to a level just above the critical point. The subcritical level consists of a pair of circles, the points removed lie on the two integral curves terminating at $(1,0,0)$, and the points inserted lie on the two integral curves emanating from $(1,0,0)$, to give a circle as the supercritical level. One can similarly analyze the critical point $(-1,0,0)$.

(iv) $m = 1$, $n = 2$. In this case we require an embedding e of $S^0 \times \text{Int } B^2$ in M^2. The surgery removes $e(S^0 \times \{0\})$, which is a pair of points, and replaces it by $\{0\} \times S^1$, i.e., by a circle. Fig. 50 illustrates how the sewing takes place in the general situation, i.e., it illustrates the diffeomorphism α between $S^0 \times (\text{Int } B^2 - \{0\})$ and $(\text{Int } B^1 - \{0\}) \times S^1$.

150   *Differential Topology*

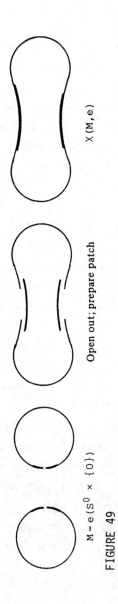

FIGURE 49

11. Surgery

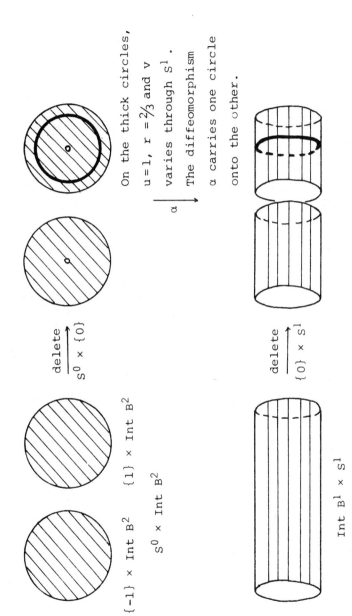

FIGURE 50

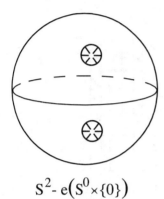

$S^2 - e(S^0 \times \{0\})$

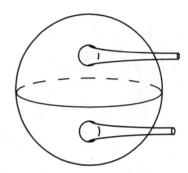

Still $S^2 - e(S^0 \times \{0\})$ but with $e(S^0 \times (\text{Int } B^2 - \{0\}))$ stretched, the holes left by removing $e(S^0 \times \{0\})$ are enlarged.

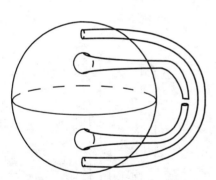

Bend the tubes $e(S^0 \times (\text{Int } B^2 - \{0\}))$ a bit; lay the patch Int $B^1 \times S^1$ alongside.

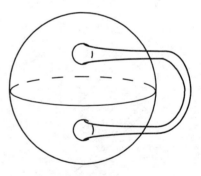

$\chi(S^2, e) \approx T^2$

FIGURE 51

11. Surgery

Now suppose that $M^2 = S^2$, so that $e(S^0 \times \text{Int } B^2)$ consists of a pair of disks on the sphere. To obtain $X(S^2,e)$, the surgery removes a pair of points and sews in a circle. Fig. 51 shows how this surgery may be performed. Note that the resulting manifold is (diffeomorphically!) T^2.

The reader should verify that if N is obtainable from M by a surgery to type (m,n) then M is obtainable from N by a surgery of type (n,m)—Exercise 2. In addition, the reader should carry out a few more specific surgeries.

Example iii alludes to the relationship between surgery and the crossing of a critical level. If this level contains only one critical point, then the supercritical level is always obtainable from the subcritical level by a surgery, the type of which is determined by the index of the critical point. However, as in Theorem 10.2, we can say more. We can determine the structure of the manifold between the two levels. This involves the concept of trace, which is studied in the next chapter.

EXERCISES

1. Give an example of two manifolds with boundary, M and N, for which $\partial M = \partial N$ ($\neq S^m \times S^n$). Under what conditions can a (smooth) surgery involving replacement of one of these manifolds by the other be performed?
2. Suppose N is obtainable from M by a surgery of type (m,n). Prove that M is obtainable from N by a surgery of type (n,m).
3. Put limerick 1.II into the context of the adjunction manifolds of this chapter.

12
THE TRACE OF A SURGERY

In this chapter we show how to construct a manifold with boundary associated with a surgery. Suppose we have a type (m,n) surgery which transforms the manifold M into the manifold $\chi(M,e)$. Then we will construct a manifold with boundary $\omega(M,e)$, called the trace of the surgery, with the property that the boundary consists (up to diffeomorphism, in a natural manner) of two pieces, M and $\chi(M,e)$ [thus $\omega(M,e)$ has dimension m + n]. Having done this we will then relate it to Morse functions. We will find in Theorem 1 that there is a natural Morse function defined on $\omega(M,e)$ and having a single critical point, such that M is a subcritical level and $\chi(M,e)$ a supercritical level. Theorem 2, on the other hand, will show that if there is a single critical point of a Morse function between two particular regular levels, then the portion of the manifold between the regular levels is diffeomorphic to the trace of a naturally determined surgery, the type determined by the index of the critical point. Thus studying the structure of a manifold in the vicinity of a critical level is equivalent to studying the trace of a surgery. These alternative viewpoints will complement one another in Chaps. 13 and 14.

As in our study of a manifold between two regular levels with no intervening critical level, integral curves play an important role, and we will motivate our study of the trace by looking at an example involving the use of integral curves.

Example Consider our standing example $f : T^2 \to \mathbb{R}$ given by $f(x,y,z) = x$. The point $(1,0,0)$ is the only critical point between the two regular levels $f^{-1}(0)$ and $f^{-1}(2)$. Let ξ be a gradientlike vector field for f, for example, that constructed in Chap. 10 and depicted in Fig. 40. Notice that two integral curves from $f^{-1}(0)$ terminate at $(1,0,0)$ and two integral curves emanate from $(1,0,0)$. All remaining integral curves from $f^{-1}(0)$ climb to the level $f^{-1}(2)$. Let p,q denote the points of $f^{-1}(0)$ the integral curves from which terminate at $(1,0,0)$ (in the case of the gradientlike vector field constructed in Chap. 10, $\{p,q\} = \{(0,1,0), (0,-1,0)\}$). Then, as in Theorem 10.2, if we remove from $f^{-1}([0,2])$ the integral curves terminating at or emanating from $(1,0,0)$, the result is naturally diffeomorphic (by use of the remaining integral curves) to $[f^{-1}(0) - \{p,q\}] \times [0,2]$. Notice that the first factor of this product is the level $f^{-1}(0)$ with a 0-sphere removed, which is the first step in carrying out a surgery of type (1,1). We could try to recreate $f^{-1}([0,2])$ by adding a patch to this product. What is the form of the patch? As in Chap. 11, it should consist of more than just the missing integral curves. In fact, we could take, in addition to the missing integral curves, all integral curves which begin at $f^{-1}(0)$ within, say, 1 unit of p or q; the union of these curves forms a neighborhood of the missing curves. This neighborhood could now be patched onto the product by the natural (integral curve induced) diffeomorphism. In Fig. 52, this patch is shown on $f^{-1}([0,2])$ and separately, flattened out.

According to Morse's theorem 9.4, there is a chart (U,φ) about $(1,0,0)$ so that

$$f\varphi^{-1}(x,y) = -x^2 + y^2 + 1 \qquad \text{for every } (x,y) \in \varphi(U)$$

The flattened patch in Fig. 52 is, effectively, $\varphi(U)$, with the x and y axes in their usual directions. That portion of $\varphi(U)$ which is patched onto $f^{-1}(0)$ satisfies $-x^2 + y^2 = -1$, that which is patched onto $f^{-1}(2)$ satisfies $-x^2 + y^2 = 1$, and the rest satisfies

$$-1 \leq -x^2 + y^2 \leq 1$$

12. The Trace of a Surgery

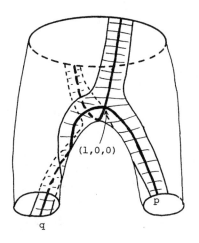
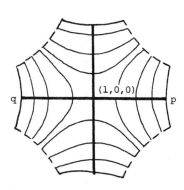

FIGURE 52

By Exercise 10.2, the width of the neighborhood is determined by an inequality of the form $|xy| <$ constant, where the constant is in effect the width. Thus the patch is determined by the inequalities

$$-1 \leq -x^2 + y^2 \leq 1 \qquad |xy| < \text{constant}$$

For convenience later, we will take the constant to be the number sinh 1 cosh 1 .

The situation described in the example above immediately generalizes to any critical point of any index, except that there should be no other critical points between the levels 1 unit on either side of the critical level (if there were, then the constant and/or ±1 would need to be decreased in the inequalities).

There is a further, minor, technical point. The definition of adjunction manifold in Chap. 11 needs to be generalized to manifolds with boundary. The generalization is obvious.

We require the following manifold with boundary: it will be the patch referred to in the example above. Let

$$P_{m,n} = \{(x,y) \in \mathbb{R}^m \times \mathbb{R}^n \mid -1 \leq -|x|^2 + |y|^2 \leq 1 \text{ and}$$
$$|x||y| < \sinh 1 \cosh 1\}$$

$P_{m,n}$ is an $(m + n)$-manifold with boundary, $\partial P_{m,n}$ consisting of two pieces. One portion, $\partial_{-1}P_{m,n}$, is defined by $-|x|^2 + |y|^2 = -1$, and the other, $\partial_1 P_{m,n}$, is defined by $-|x|^2 + |y|^2 = 1$. Note that $\partial_{-1}P_{m,n}$ is diffeomorphic to $S^{m-1} \times \text{Int } B^n$ by the diffeomorphism

$$(u \cosh r, v \sinh r) \longleftrightarrow (u, rv) \qquad u \in S^{m-1}, v \in S^{n-1},$$
$$0 \leq r < 1$$

and that $\partial_1 P_{m,n}$ is diffeomorphic to $\text{Int } B^m \times S^{n-1}$ by the diffeomorphism

$$(u \sinh r, v \cosh r) \longleftrightarrow (ru, v) \qquad u \in S^{m-1}, v \in S^{n-1},$$
$$0 \leq r < 1$$

Fig. 53 shows $P_{m,n}$ for $(m,n) = (0,2), (1,1), (2,0),$ and $(1,2)$. Note that $P_{1,1}$ is the patch pictured in Fig. 52, and had we drawn analogues of Fig. 52 for the critical points $(-3,0,0)$ and $(3,0,0)$ we would have drawn $P_{0,2}$ and $P_{2,0}$, respectively.

Note that $P_{0,3}$ and $P_{3,0}$ are just B^3, and $P_{1,2}$ and $P_{2,1}$ are obtained by revolving $P_{1,1}$ about the y axis or the x axis (respectively). Thus $P_{2,1}$ looks just like $P_{1,2}$ but with the x axis relabeled the y axis and the y_1 and y_2 axes relabeled the x_1 and x_2 axes.

As has already been suggested, we now should describe the natural integral curves on the patch $P_{m,n}$. These curves (cf. Theorem 10.2 and Exercise 10.2) satisfy the differential equations $(dx/dt, dy/dt) = (-x,y)$, the solutions being $(x,y) = (ae^{-t}, be^t)$. For any point $p \in \partial_{-1}P_{m,n}$, let γ_p denote the integral curve from p. If we set the boundary condition by decreeing that $\gamma_p(0) = p$, and if p is the point $(u \cosh r, v \sinh r)$, then

$$\gamma_p(t) = (e^{-t} u \cosh r, e^t v \sinh r)$$

12. The Trace of a Surgery

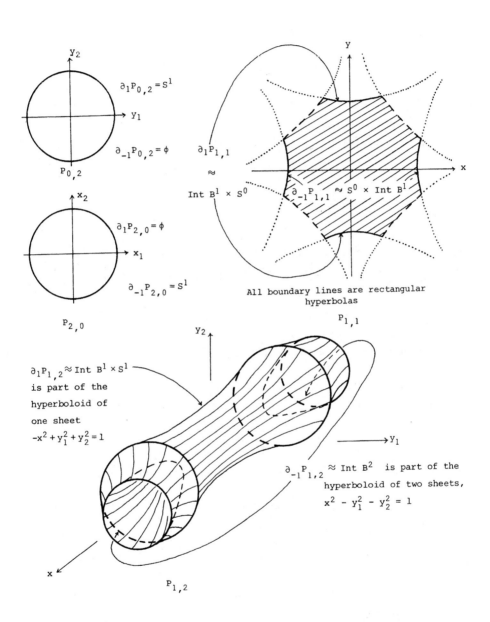

FIGURE 53

Note that

$$\gamma_p(\ln \coth r) = (u \sinh r, v \cosh r) \in \partial_1 P_{m,n}$$

except when $r = 0$. Integral curves starting from $(u,0)$ end at $(0,0)$ and these are the only curves ending at $(0,0)$. Integral curves ending at $(0,v)$ emanate from $(0,0)$ and these are the only such curves.

We now use these integral curves to obtain a diffeomorphism related to the diffeomorphism α of Chap. 11. Define the diffeomorphism

$$\beta : S^{m-1} \times (\text{Int } B^n - \{0\}) \times [-1,1] \to P_{m,n} - (\mathbb{R}^m \times \{0\}) - (\{0\} \times \mathbb{R}^n)$$

by $\beta(u,rv,t) = (x,y)$, where (x,y) is the unique point on the curve $\gamma_{(u \cosh r, v \sinh r)}$ defined above and satisfying $-|x|^2 + |y|^2 = t$. When $t = -1$, we obtain $\beta(u,rv,-1) = (u \cosh r, v \sinh r)$, so β extends the diffeomorphism between $\partial_{-1} P_{m,n}$ and $S^{m-1} \times \text{Int } B^n$. When $t = 1$, we obtain $\beta(u,rv,1) = (u \sinh r, v \cosh r)$, so β extends the diffeomorphism between $\partial_1 P_{m,n}$ and $\text{Int } B^m \times S^{n-1}$.

Definition Let M^{m+n-1} be a manifold and $e : S^{m-1} \times \text{Int } B^n \to M$ be an embedding. Let $\omega(M,e)$ denote the adjunction manifold

$$[M - e(S^{m-1} \times \{0\})] \times [-1,1] \cup_{\beta(e^{-1} \times 1)} P_{m,n}$$

Then $\omega(M,e)$ is called the *trace* of the surgery determined by e.

Note that $\omega(M,e)$ is a manifold with boundary. Its boundary comes from two sources:

$$\partial([M - e(S^{m-1} \times \{0\})] \times [-1,1]) = [M - e(S^{m-1} \times \{0\})] \times \{-1,1\}$$

and $\partial P_{m,n} = \partial_{-1} P_{m,n} \cup \partial_1 P_{m,n}$. The adjunction $\beta(e^{-1} \times 1)$ sews $\partial_{-1} P_{m,n} \approx S^{m-1} \times \text{Int } B^n$ to $[M - e(S^{m-1} \times \{0\})] \times \{-1\}$ to give us $\partial_{-1}\omega(M,e)$, which is effectively M, while it sews $\partial_1 P_{m,n} \approx \text{Int } B^m \times S^{n-1}$ to $[M - e(S^{m-1} \times \{0\})] \times \{1\}$ to give us $\partial_1\omega(M,e)$, which is effectively $\chi(M,e)$.

12. The Trace of a Surgery

Thus the trace of a surgery is a manifold with boundary with the property that one part of the boundary is the original manifold and the other part of the boundary is the surgically altered manifold (up to diffeomorphism).

Examples (i) $m = 0$. As already noted in Chap. 11, surgery of type $(0,n)$ adds a disjoint, far, sphere of dimension $n - 1$ to the otherwise unchanged manifold. Again, the embedding $\beta(e^{-1} \times 1)$ has empty domain and range, and $P_{0,n}$ is an n-ball bounded by the added sphere. Thus the trace of the surgery is just the product of the original manifold with $[-1,1]$ together with a disjoint n-ball. If we thicken the torus of Fig. 48, making it into a circular pipe, and fill in the sphere, we will get the required trace. One part of the boundary is the original torus, say, the outer boundary of the pipe, and the other part of the boundary is the sphere, i.e., the boundary of the ball, together with another torus, say, the inner boundary of the pipe.

(ii) $n = 0$. As noted in Chap. 11, there needs to be a submanifold of M^{m-1} diffeomorphic to S^{m-1} to permit such a surgery. The trace is the product of what is left of M after the $(m - 1)$-sphere on which the surgery is being performed is removed, with the interval $[-1,1]$ together with a disjoint ball which is bounded by the sphere.

(iii) $m = n = 1$. Consider the surgery of type $(1,1)$ depicted in Fig. 49. Then $[M - e(S^0 \times \{0\})] \times [-1,1]$ looks like a pair of cylinders each with a longitudinal slit. We need to describe how $P_{1,1}$ is sewn on by the adjunction diffeomorphism $\beta(e^{-1} \times 1)$. According to the definition of β, $\partial_{-1} P_{1,1}$ is sewn to the -1 end of the cylinders and $\partial_1 P_{1,1}$ to the 1 end. Thus to perform the adjunction we should raise $\partial_1 P_{1,1}$ up to the 1 level and lower $\partial_{-1} P_{1,1}$ to the -1 level. In general, the point $(x,y) \in P_{1,1}$ should be raised to the level $-x^2 + y^2$ prior to sewing to the cylinders. The integral curves, which are parts of rectangular hyperbolas on $P_{1,1}$ should be straightened out. This deformation of $P_{1,1}$ gives rise to a saddle-shaped surface [$(1,1)$ surgery takes place at a saddle point] which we then sew to the cylinders to give us the trace, a pair of trousers depicted in Fig. 54.

162   Differential Topology

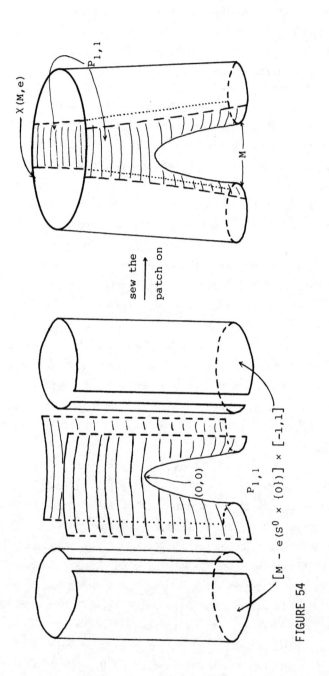

FIGURE 54

12. The Trace of a Surgery   163

(iv) $m = 1$, $n = 2$. Consider the surgery of type $(1,2)$ depicted in Fig. 51. Since $S^2 \times [-1,1]$ is a thickened sphere, naturally diffeomorphic to the set of all points of $\mathbb{R}^3$ distant at least 1 and at most 3 from the origin ($S^2 \times \{t\}$ corresponding to the sphere of radius $t + 2$), the portion $[S^2 - e(S^0 \times \{0\})] \times [-1,1)$ of $\omega(S^2,e)$ looks like the above set with two of the radial lines removed. The patch $P_{1,2}$ is depicted in Fig. 53. To sew $P_{1,2}$ to the thickened sphere, we more or less follow the sequence of pictures in Fig. 51: $\partial_{-1} P_{1,2}$ is sewn to $S^2 - e(S^0 \times \{0\})$ by e, each piece of $\partial_{-1} P_{1,2}$ covering one of the two holes, and higher levels of $P_{1,2}$ (i.e., subsets where $x^2 - y_1^2 - y_2^2$ is constant) are sewn to the deformed spheres. In the end we will obtain a manifold with boundary which, from the outside, (i.e., whose outer boundary) looks like the last picture of Fig. 51. Its inner boundary will be S^2, and in between there will be a thickened sphere with two missing radial lines together with the solid handle, which, in effect, came from $P_{1,2}$.

THEOREM 1. If $\chi(M,e)$ is obtained from M by surgery of type (m,n), then there is a Morse function $g : \omega(M,e) \to [-1,1] \subset \mathbb{R}$ such that $g^{-1}(-1) = M$, $g^{-1}(1) = \chi(M,e)$, and g has exactly one critical point, the index of which is m.

Proof: Using the notation of the definition of trace, define g by

$g(q,t) = t$ if $(q,t) \in [M - e(S^{m-1} \times \{0\})] \times [-1,1]$

$g(x,y) = -|x|^2 + |y|^2$ if $(x,y) \in P_{m,n}$

We have slightly abused notation here. Strictly, $[M - e(S^{m-1} \times \{0\})] \times [-1,1]$ and $P_{m,n}$ are not subsets of $\omega(M,e)$, so that (q,t) and (x,y) do not really represent typical elements of $\omega(M,e)$. However, in our definition of adjunction manifold we showed how the two pieces which are sewn together may be naturally thought of as subsets (using inclusion functions i and j). Thus we should really write $gi(q,t) = t$ and $gj(x,y) = -|x|^2 + |y|^2$, where i and j are the inclusion functions.

It is clear now that g is well defined because the map β used in defining the adjunction locates the point (x,y) on a certain curve so that $-|x|^2 + |y|^2 = t$; thus if (q,t) and (x,y) represent the same point of ω(M,e) then $g(q,t) = t = -|x|^2 + |y|^2 = g(x,y)$. The smoothness of g follows from the fact that it is smooth on each of the two open subsets of ω(M,e) mentioned in the definition. The rank of g at points of $[M - e(S^{m-1} \times \{0\})] \times [-1,1]$ is clearly 1 ; one merely calculates the jacobian of g with respect to a product chart, in which case the last entry must be dt/dt = 1. On the other hand, $(P_{m,n},1)$ provides, essentially, a chart covering the rest of the manifold. With respect to this chart, the function g is already in the form of Morse's theorem, thus showing that (0,0) is the unique critical point, that it is nondegenerate, and that the index is m. □

We now derive a converse of Theorem 1: if the region between two regular levels of a Morse function contains only one critical point, say of index λ , then it is diffeomorphic to the trace of a surgery whose type depends only on λ and the dimension of the manifold. The standing example discussed at the beginning of this chapter motivates our construction. Surgery is performed on the lower level, the sphere to be removed from this level consisting of all points lying on integral curves which terminate at the critical point. The sphere to be sewn in emanates from the critical point.

Suppose M^m is a manifold and $f : M \to \mathbb{R}$ is a Morse function having only one critical point $p \in f^{-1}([-1,1])$. Assume that f(p) = 0 and that the index of p is λ . Let ξ be a gradientlike vector field for f . Then there is a chart (U,φ) about p such that

$$f\varphi^{-1}(x,y) = -|x|^2 + |y|^2$$

and the components of $\xi(\varphi^{-1}(x,y))$ with respect to (U,φ) are (-x,y) , for all $(x,y) \in \varphi(U)$. We may also assume that φ(U) is large enough so that the embeddings e and $\tilde{e}$ below are well defined. This assumption is made solely to simplify the algebra. The reader might like to see how it can be removed.

12. The Trace of a Surgery

Definitions The *lower characteristic embedding*,

$$\underset{\sim}{e} : S^{\lambda-1} \times \text{Int } B^{m-\lambda} \longrightarrow M_{-1}$$

is defined by

$$\underset{\sim}{e}(u, rv) = \varphi^{-1}(u \cosh r, v \sinh r)$$

where, as for $P_{\lambda, m-\lambda}$, we take $u \in S^{\lambda-1}$, $v \in S^{m-\lambda-1}$, and $r \in [0,1)$. The set $\underset{\sim}{e}(S^{\lambda-1} \times \{0\})$ is called the *lower sphere* of p in M_{-1}; it is diffeomorphic to $S^{\lambda-1}$. The lower characteristic embedding and lower sphere may be defined on any other level below 0 until the level is so low that some other critical point intervenes, by sliding along integral curves.

Similarly one defines the *upper characteristic embedding*

$$\tilde{e} : \text{Int } B^{\lambda} \times S^{m-\lambda-1} \longrightarrow M_1$$

by $\tilde{e}(ru, v) = \varphi^{-1}(u \sinh r, v \cosh r)$, and the *upper sphere* of p to be $\tilde{e}(\{0\} \times S^{m-\lambda-1})$.

Note that the lower and upper characteristic embeddings are essentially just the diffeomorphisms between $S^{\lambda-1} \times \text{Int } B^{m-\lambda}$ (respectively Int $B^{\lambda} \times S^{m-\lambda-1}$) and $\partial_{-1} P_{\lambda, m-\lambda}$ (respectively $\partial_1 P_{\lambda, m-\lambda}$). Moreover, by analogy with the integral curves on $P_{\lambda, m-\lambda}$, we have that any integral curve for ξ passing through M_{-1} terminates at p iff it passes through the lower sphere, and any integral curve for ξ passing through M_1 emanates from p iff it passes through the upper sphere. Thus the change in levels as we pass through p is to squeeze the lower sphere to the point p, then expand p into the upper sphere. Fig. 55 shows some examples, where the Morse functions is height up the page.

THEOREM 2. *Let* $f : M^m \longrightarrow \mathbb{R}$ *be a Morse function on a compact manifold with only one critical point, say p, in* $f^{-1}([-1,1])$. *Suppose that* $f(p) = 0$ *and that the index is* λ. *Denote by*

$$\underset{\sim}{e} : S^{\lambda-1} \times \text{Int } B^{m-1} \longrightarrow M_{-1}$$

the lower characteristic embedding with

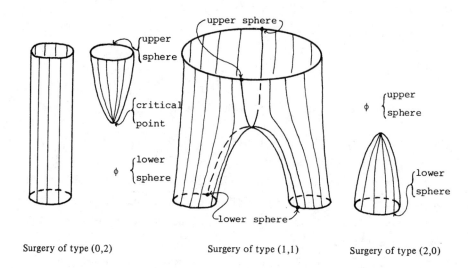

Surgery of type (0,2)   Surgery of type (1,1)   Surgery of type (2,0)

FIGURE 55

respect to some chart. Then there is a diffeomorphism

$$h : \omega(M_{-1}, \underset{\sim}{e}) \to f^{-1}([-1,1])$$

such that $h(\partial_{-1}\omega(M_{-1},\underset{\sim}{e})) = M_{-1}$, $h(\partial_1 \omega(M_{-1},\underset{\sim}{e})) = M_1$, and if $g : \omega(M_{-1},\underset{\sim}{e}) \to \mathbb{R}$ is the Morse function constructed in Theorem 1, then

$$fh(x) = g(x) \quad \text{for every } x \in \omega(M_{-1},\underset{\sim}{e})$$

Proof: As in the proof of Theorem 1 we will slightly abuse the notation. Using also the notation set up before the theorem, if $(q,t) \in [M_{-1} - \underset{\sim}{e}(S^{\lambda-1} \times \{0\})] \times [-1,1]$, let $h(q,t)$ be that point of $f^{-1}[-1,1]$ lying on the integral curve through q for which $fh(q,t) = t$; if $(x,y) \in P_{\lambda,m-\lambda}$, set $h(x,y) = \varphi^{-1}(x,y)$. One can check that h satisfies the requirements of the theorem. □

One can adjust the conclusion of Theorem 2 in the case where the interval $[-1,1]$ in the inverse image of which f has only one

12. The Trace of a Surgery

critical point is replaced by a more general interval. Suppose that $f : M^m \to \mathbb{R}$ is a Morse function with only one critical point, say p, in $f^{-1}([a,b])$ with $f(p) = (a + b)/2$, and the index still is λ. Then there is a diffeomorphism

$$h : \omega(M_a, \underset{\sim}{e}) \to f^{-1}([a,b])$$

with $h(\partial_{-1} \omega(M_a, \underset{\sim}{e})) = M_a$, $h(\partial_1 \omega(M_a, \underset{\sim}{e})) = M_b$, and if $g : \omega(M_a, \underset{\sim}{e}) \to \mathbb{R}$ is as before, then

$$fh(x) = \varepsilon g(x) \qquad \text{for every } x \in \omega(M_a, \underset{\sim}{e})$$

where $\varepsilon : [-1,1] \to [a,b]$ is the linear diffeomorphism given by $\varepsilon(t) = (b - a)t/2 + (a + b)/2$.

The results of this chapter tell us that to determine the structure of a manifold given a Morse function on it, we perform a sequence of surgeries, one corresponding to each critical point of the function, each surgery of a type determined by the index of the critical point. The manifold is the trace of all of these surgeries, with extra connecting links given by Theorem 10.2 where necessary. If two critical points lie on the same level, then we can use Exercise 9.5 to adjust the Morse function slightly so that this is no longer the case. Such an adjustment ensures that it is not necessary to perform two surgeries simultaneously.

EXERCISE

1. Complete the proof of Theorem 2, i.e., verify that h is a diffeomorphism satisfying all of the requirements.

13
SURGERY ON A SURFACE

Definition A *surface* is a compact connected 2-manifold.

Examples K^2, P^2, S^2, and T^2 are surfaces but B^2 and $\mathbb{R}^2$ are not, according to our definition.

Our ultimate goal, achieved in Chap. 14, is to classify orientable surfaces. Motivated by the concluding paragraph in Chap. 12, we study the traces of surgeries of type $(\lambda, 2 - \lambda)$, since these are put together to form a surface, making sure that we identify all possibilities. We find that there are five basic kinds, one of which is not orientable. The other four are exhibited by our standing example $f : T^2 \to \mathbb{R}$.

Before analyzing surgeries of type $(\lambda, 2 - \lambda)$, we need to determine all possible compact 1-manifolds, since these are the levels on which the surgeries will be performed.

PROPOSITION 1. Up to diffeomorphism, S^1 is the only compact connected 1-manifold.

Proof: Suppose M is a compact connected 1-manifold. By Theorem 9.6, there is a Morse function $f : M \to \mathbb{R}$. The critical points of f must have indices 0 or 1, i.e., must be minima or maxima. Note that by the Heine-Borel theorem (3.2) and Theorem 3.3, f must have at least one minimum and at least one maximum. Using prototype Lemma 4.6, we may assume that if p is a minimum and q is a maximum then $f(p) < f(q)$. Thus all surgeries of type (0,1) are performed before any surgeries of type (1,0).

Suppose there are n surgeries of type (0,1) . The trace of one such surgery is a U shape, which can be embedded in the plane; thus we can draw an accurate picture of it. The trace of n surgeries of type (0,1) will be n such U shapes, of which we can draw a picture, with height up the page as the Morse function. In effect, by drawing a picture, we are choosing an embedding of the trace in the plane, as in Fig. 56.

Consider the first surgery of type (1,0) on this. It involves removal of a sphere S^0 , i.e., of two points, the trace being a ∩ shape. If the two points lie in the same U , then this will be closed off, giving a disconnection of M , unless n = 1 . If n = 1 , then by closing off the U , we get S^1 and the proof is complete. Otherwise the first (1,0) surgery connects together two different U shapes. We might as well assume that they are adjacent U shapes, for if not, then we should choose a different embedding, i.e., redraw Fig. 56, so that this is the case. The same applies to the next n - 2 type (1,0) surgeries. Having performed these n - 1 type (1,0) surgeries, our trace is as in Fig. 57. This is not a manifold, so there must be another type (1,0) surgery which, of necessity, joins the two remaining ends, giving us S^1 up to diffeomorphism. □

(n U's)

FIGURE 56

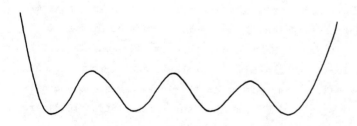

FIGURE 57

13. Surgery on a Surface

COROLLARY 2. Up to diffeomorphism, the only compact 1-manifolds are those consisting of finitely many disjoint circles.

We are now in a position to analyze surgeries of types $(\lambda, 2 - \lambda)$. Note that there are three types: $(0,2)$, $(1,1)$, $(2,0)$. We look at the three types separately.

Surgery of Type (0,2) This results in the removal of a sphere of dimension $0 - 1$, i.e., of ϕ, and its replacement by a sphere of dimension $2 - 1$, i.e., by S^1. The trace of such a surgery performed on M is $(M \times [-1,1]) + B^2$, as already noted in Chap. 12; + denotes the disjoint sum.

Surgery of Type (2,0) This results in the removal of S^1 and its replacement by ϕ. Since this surgery is to be performed on a compact 1-manifold, which must be the sum of finitely many circles, its effect is to cap off one of these circles.

Surgery of Type (1,1) This results in the removal of a 0-sphere and its replacement by another 0-sphere. We will find that there are three distinct kinds of (1,1) surgery, which we will call connecting, disconnecting, and twisting.

Suppose that M is a compact 1-manifold, i.e., a finite sum of circles. A surgery of type (1,1) requires an embedding of $S^0 \times \text{Int } B^1$ in M and the sewing on of the patch $\text{Int } B^1 \times S^0$. Of course we want to determine all possible traces, but we begin by looking at the surgery itself. Fig. 58 shows how the two sets $S^0 \times \text{Int } B^1$ and $\text{Int } B^1 \times S^0$ are related, as well as the diffeomorphism α used in defining $\chi(M,e)$ in Chap. 11.

Now suppose given an embedding $e : S^0 \times \text{Int } B^1 \to M$. Using Fig. 58 to guide us and choosing an embedding of M in $\mathbb{R}^3$ in a careful manner, we can draw a picture illustrating how to get from M to $\chi(M,e)$. Fig. 59 shows a first step. In this we have shown only part of M near $e(S^0 \times \text{Int } B^1)$, and this part is so embedded in $\mathbb{R}^3$ that it actually lies in $\mathbb{R}^2$ in such a way that $e(\{-1\} \times \text{Int } B^1)$

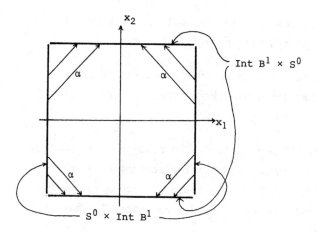

FIGURE 58

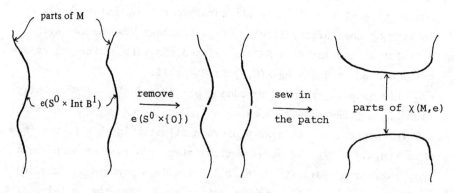

FIGURE 59

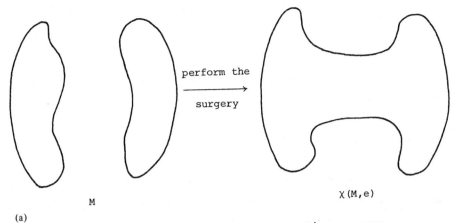

(a) two circles become one; a connecting surgery

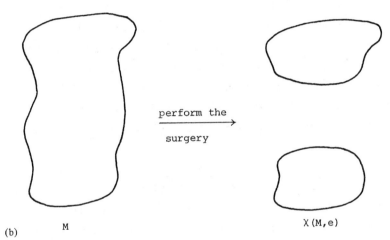

(b) one circle splits into two; a disconnecting surgery

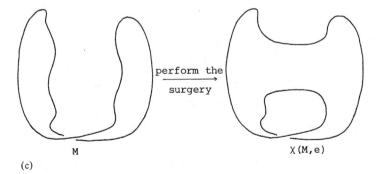

(c) one circle is transformed into another; a twisting surgery

FIGURE 60

lies in the left part and $e(\{1\} \times \text{Int } B^1)$ lies in the right part, and that in each case the upward direction on the page denotes the positive direction on the Int B^1 factor of $e(S^0 \times \text{Int } B^1)$. That we can choose such an embedding of M in $\mathbb{R}^3$ should be clear. The advantage of such a choice is that it is easy to see how to carry out the surgery using the sewing map illustrated in Fig. 58.

To complete the picture of the surgery from M to $\chi(M,e)$, we must fill in the rest of M, this part surviving the surgery unscathed. By Corollary 2, M consists of several circles. Since the surgery affects only the circle or two circles containing $e(S^0 \times \text{Int } B^1)$, we might as well ignore any other circles lying in M. Thus, by joining up the four free ends of the parts of M shown in Fig. 59 in all possible ways (up to diffeomorphism) so as to form one or two circles, we will determine all possible kinds of surgery of type (1,1). Fig. 60 depicts the possibilities.

Notice that the three possibilities illustrated by Fig. 60 are mutually distinct since for any two either M or $\chi(M,e)$ are not diffeomorphic.

In the first case, $e(\{-1\} \times \text{Int } B^1)$ and $e(\{1\} \times \text{Int } B^1)$ lie in different components of M and the surgery joins these two components to form a single circle, a situation we have already met (cf. Figs 49 and 54). As already shown in Fig. 54, the trace is a pair of trousers. This surgery is called the *connecting* surgery of type (1,1).

In the second case, $e(\{-1\} \times \text{Int } B^1)$ and $e(\{1\} \times \text{Int } B^1)$ lie in the same component of M but their natural directions disagree within that component (i.e., one points in the clockwise direction and the other in the counterclockwise direction). This surgery splits the circle into two circles, so is called the *disconnecting* surgery of type (1,1). Its trace is a pair of trousers upside down and it is the reverse of the connecting surgery. It is typically represented by the embedding

$$S^0 \times \text{Int } B^1 \to S^1$$

defined by

13. Surgery on a Surface

$$(\pm 1, r) \mapsto (r, \pm\sqrt{1 - r^2})$$

If $e : S^0 \times \text{Int } B^1 \mapsto S^1$ is any embedding, then $\omega(S^1, e)$ will be diffeomorphic to the trace of the above typical disconnecting surgery iff $e(-1,-\frac{1}{2})$, $e(-1,\frac{1}{2})$, $e(1,\frac{1}{2})$, $e(1,-\frac{1}{2})$ lie in order around the circle.

In the third case, $e(\{-1\} \times \text{Int } B^1)$ and $e(\{1\} \times \text{Int } B^1)$ lie in the same component of M and their natural directions agree within that component. This surgery is typically represented by the embedding

$$S^0 \times \text{Int } B^1 \to S^1$$

defined by

$$(\pm 1, r) \mapsto \pm(r, \sqrt{1 - r^2})$$

If one performs this particular surgery the effect is to twist the circle, and the surgery is called the *twisting* surgery of type $(1,1)$. If $e : S^0 \times \text{Int } B^1 \to S^1$ is any embedding, then $\omega(S^1, e)$ will be diffeomorphic to the trace of the above typical twisting surgery iff $e(-1,-\frac{1}{2})$, $e(-1,\frac{1}{2})$, $e(1,-\frac{1}{2})$, $e(1,\frac{1}{2})$ lie in order around the circle.

Our standard Morse function on the torus illustrates the first four surgeries considered above. The twisting surgery we have also met already, viz., in Chap. 6. Fig. 27 depicts the projective plane as the trace of three surgeries: a $(0,2)$ surgery, a twisting surgery of type $(1,1)$, and a $(2,0)$ surgery. In fact no other compact manifold is the trace of exactly three surgeries.

The following lemma tells us that we may ignore the twisting surgery if we are interested in classifying orientable surfaces only.

LEMMA 3. *The trace of a twisting surgery of type $(1,1)$ is non-orientable.*

Proof: We take the standard embedding $e : S^0 \times \text{Int } B^1 \to S^1$ defined by $e(\pm 1, r) = \pm(r, \sqrt{1 - r^2})$ and show that $\omega(S^1, e)$ is non-orientable. By Theorem 6.3, it suffices to find two charts (U, φ)

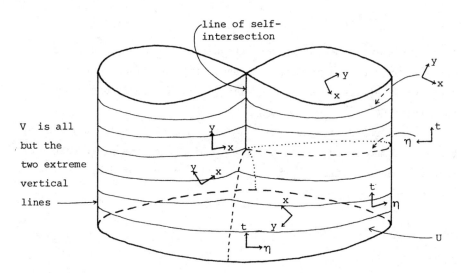

FIGURE 61

and (V,ψ) in the structure of $\omega(S^1,e)$ such that U and V are both connected but that $\Delta(\varphi\psi^{-1})$ changes sign in $\psi(U \cap V)$. Let

$$U = i(\{(\xi,\eta) \in S^1 \mid \xi > 0\} \times (-1,0)) \qquad \varphi i((\xi,\eta),t) = (\eta,t)$$
$$V = j(P_{1,1}) \qquad \psi = j^{-1}$$

where $i : [S^1 - e(S^0 \times \{0\})] \times [-1,1] \to \omega(S^1,e)$ and $j : P_{1,1} \to \omega(S^1,e)$ are the standard inclusions into the adjunction manifold; cf. Fig. 61. Note that $U \cap V$ consists of two pieces:

$$A^+ = i(\{(\xi,\eta) \in S^1 \mid \xi > 0, \eta > 0\} \times (-1,0))$$
$$A^- = i(\{(\xi,\eta) \in S^1 \mid \xi > 0, \eta < 0\} \times (-1,0))$$

Recall (cf. the proof of Lemma 11.1) that $\beta(e^{-1} \times 1) = j^{-1}i$, so that

$$\psi(A^+) = \beta(e^{-1} \times 1)(\{(\xi,\eta) \in S^1 \mid \xi > 0, \eta > 0\} \times (-1,0))$$

Now if $(\xi,\eta) \in S^1$ is such that $\xi > 0$ and $\eta > 0$, then for some $r > 0$, $(\xi,\eta) = e(1,r)$, so that by definition, $\beta(e^{-1} \times 1)((\xi,\eta),t)$ lies on the integral curve in $P_{1,1}$ starting at $(\cosh r, \sinh r)$. Therefore

13. Surgery on a Surface

$\beta(e^{-1} \times 1)((\xi,\eta),t)$ lies in the quadrant $x > 0$, $y > 0$, of $P_{1,1}$. Thus $\psi(A^+)$ lies in this quadrant of $P_{1,1}$. Moreover, if $(x,y) \in \psi(A^+)$, then

$$\varphi\psi^{-1}(x,y) = (\eta,t)$$

where, by definition of β, $t = -x^2 + y^2$ and (x,y) lies on the integral curve starting at $(\cosh r, \sinh r)$. Thus $xy = \cosh r \sinh r = \tfrac{1}{2} \sinh 2r$. Summarizing, for $(x,y) \in \psi(A^+)$

$$\varphi\psi^{-1}(x,y) = (\sqrt{1-r^2},\, -x^2 + y^2)$$

where $xy = \tfrac{1}{2} \sinh 2r$.

We can now form the jacobian matrix

$$D(\varphi\psi^{-1})(x,y) = \begin{pmatrix} \dfrac{-r}{\sqrt{1-r^2}} \dfrac{\partial r}{\partial x} & \dfrac{-r}{\sqrt{1-r^2}} \dfrac{\partial r}{\partial y} \\ \\ -2x & 2y \end{pmatrix}$$

and hence

$$\Delta(\varphi\psi^{-1})(x,y) = \dfrac{-2r}{\sqrt{1-r^2}} \left(y \dfrac{\partial r}{\partial x} + x \dfrac{\partial r}{\partial y} \right)$$

But $xy = \tfrac{1}{2} \sinh 2r$, so, differentiating with respect to x,

$$y = \cosh 2r\, \dfrac{\partial r}{\partial x} \qquad \text{i.e.,} \qquad \dfrac{\partial r}{\partial x} = \dfrac{y}{\cosh 2r}$$

Similarly, $\partial r/\partial y = x/\cosh 2r$. Thus $\Delta(\varphi\psi^{-1})(x,y) = -2r(x^2+y^2)/\cosh 2r \sqrt{1-r^2}$ which is negative throughout $\psi(A^+)$.

Similar working shows that if $(x,y) \in \psi(A^-)$, then $x < 0$ and $y < 0$, and once again that

$$\Delta(\varphi\psi^{-1})(x,y) = \dfrac{-2r(x^2+y^2)}{\cosh 2r \sqrt{1-r^2}}$$

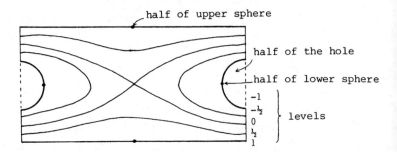

Obtain the trace by joining the two ends after twisting the rectangle

FIGURE 62

In this case, however, $r < 0$, so the transformation $\varphi\psi^{-1}$ is orientation preserving.

Remark Fig. 62 gives an alternative picture of the trace of the twisting surgery of type (1,1), as a Möbius band with a hole in it. It is quite clear how $P_{1,1}$ sits inside the trace: $P_{1,1}$ consists of everything in the picture except two integral curves which are each represented twice by vertical line segments at either end of the picture. The octagonal shape of $P_{1,1}$ becomes evident and should be compared with the standard picture of $P_{1,1}$ given in Fig. 53. Height up the page no longer measures the Morse function of Theorem 12.1. Instead, several contour lines, i.e., levels, are shown. The Möbius band with the hole cannot be embedded in $\mathbb{R}^3$ in such a way that the last coordinate gives the indicated Morse function, although it can be so embedded in $\mathbb{R}^4$. There is an apparent anomaly between Figs. 61 and 62: in Fig. 61 the critical level is apparently a figure eight; in Fig. 62, a pair of circles crossing at one point. See Exercise 1 for a reconciliation.

THEOREM 4. There are five basic kinds of surgery of type $(\lambda, 2 - \lambda)$: one kind of type (0,2), one of type (2,0) and three of type (1,1). Further, all but one, of type (1,1), has orientable trace.

Fig. 63 illustrates the five types, height up the page being the Morse function of Theorem 12.1.

13. Surgery on a Surface

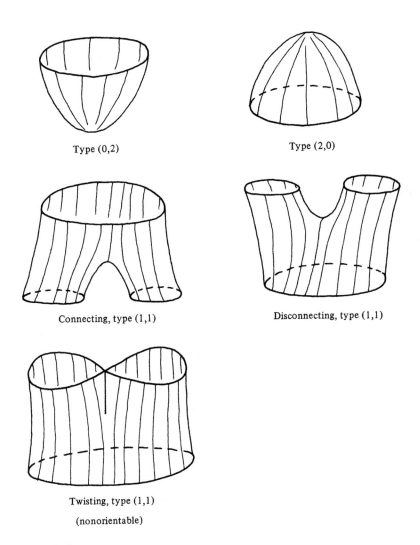

FIGURE 63

In our classification of orientable surfaces in the next chapter, we will find that such a surface can be built up by combining the four basic orientable building blocks illustrated in Fig. 63; the fifth is not used since it always gives rise to a nonorientable surface. Even using only the four orientable building blocks, we may still end up with a nonorientable surface.

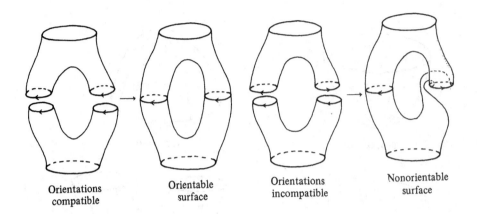

Orientations compatible   Orientable surface   Orientations incompatible   Nonorientable surface

FIGURE 64

The first two lemmas of Chap. 14 will tell us that to construct an arbitrary orientable surface, we will need only one surgery of each of types (0,2) and (2.0), and if we build up from the bottom, then we can build up all of the disconnecting type (1,1) traces before adding on any connecting type (1,1) traces. Suppose we wish to determine whether the resulting surface is orientable. We might construct an orientation for the surface as it is built up. Choose an orientation for the trace of the type (0,2) surgery. Then there will be unique compatible orientations of the traces of the disconnecting type (1,1) surgeries; thus that part, S_1, of the surface consisting of the type (0,2) trace together with the disconnecting type (1,1) traces will be orientable. Now add the first connecting type (1,1) trace to S_1. There are two possibilities: either the connecting trace is orientably compatible with S_1 or it is not and, accordingly, the extended surface is orientable or not. If each connecting type (1,1) trace is added in such a way that the orientations are compatible, then the final surface will be orientable; otherwise it will not be orientable. (See Fig. 64.)

In the next chapter, we will also find it useful to know the effects of surgeries of types (1,2) and (2,1) on a surface. We look briefly at these now.

13. Surgery on a Surface

To perform a surgery of type (2,1) on a surface we must remove a (2-1)-sphere, i.e., a circle, and replace it by a (1-1)-sphere, i.e., a pair of points. If the surface is orientable then there are two possibilities according as whether removal of the circle disconnects the surface or not. In the former case the manifold χ is not a surface according to our definition because we have insisted that our surfaces be connected. In the latter case the manifold χ is still an orientable surface.

There are two ways of performing type (1,2) surgery on S^2. One, typically represented by $e_1 : S^0 \times \text{Int } B^2 \to S^2$, where

$$e_1(t,(x,y)) = (x, y, t\sqrt{1 - x^2 - y^2})$$

results in an orientable surface: $\chi(S^2, e_1)$ is diffeomorphic to T^2 (see Fig. 65). The other type (1,2) surgery on S^2 is typically represented by $e_2 : S^0 \times \text{Int } B^2 \to S^2$, where

$$e_2(t,(x,y)) = (x, ty, t\sqrt{1 - x^2 - y^2})$$

and results in a nonorientable surface: $\chi(S^2, e_2)$ is diffeomorphic to K^2, the Klein bottle (see Fig. 66).

Note that $S^0 \times \text{Int } B^2$ consists of two copies of $\text{Int } B^2$. If we give each the orientation of $\text{Int } B^2$ induced by the natural inclusions $\text{Int } B^2 \to \{\pm 1\} \times \text{Int } B^2$, then the embedding e_1 is orientation preserving on one copy of $\text{Int } B^2$ and orientation reversing on the other, whereas e_2 is either orientation preserving on both or orientation reversing on both (depending on the orientation of S^2).

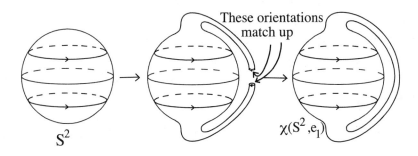

FIGURE 65

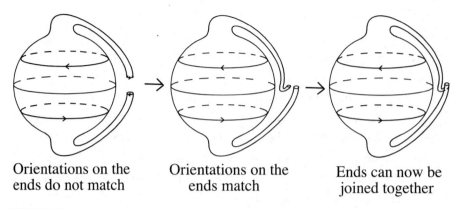

Orientations on the ends do not match   Orientations on the ends match   Ends can now be joined together

FIGURE 66

Definition An *orientable surgery* of type (1,2) on an orientable surface S is one performed along an embedding e : $S^0 \times \text{Int } B^2 \to S$, where e is orientation preserving on one copy of Int B^2 and orientation reversing on the other. Otherwise the surgery of type (1,2) is called a *nonorientable surgery*.

Remark There is only one kind of surgery of type (1,2) on a nonorientable surface (see Exercise 6). As with type (1,1) surgery, we can have a connecting surgery of type (1,2) on a 2-manifold [but a disconnecting surgery must be of type (2,1)].

LEMMA 5. Suppose S is an orientable surface and e : $S^0 \times \text{Int } B^2 \to S$ is an embedding. Then $X(S,e)$ is orientable if and only if the type (1,2) surgery performed along e is orientable.

   Proof: We use the standard notation set up in Chap. 11; thus i : $S - e(S^0 \times \{0\}) \to X(S,e)$ and j : $\text{Int } B^1 \times S^1 \to X(S,e)$ denote the natural inclusions, and α is the diffeomorphism used to define the adjunction which results in $X(S,e)$.

   Let (U,φ) be a chart on $S - e(S^0 \times \{0\})$, let $V = \text{Int } B^1 \times S^1$, and define $\psi : V \to \mathbb{R}^2$ by

$$\psi(ru,v) = (1 + ru)v \qquad \text{for every } u \in S^0,\ v \in S^1,\ r \in [0,1)$$

13. Surgery on a Surface

Note that $\psi(V) = \{(x,y) \in \mathbb{R}^2 \mid 0 < |x| < 2\}$, with the 1 end of the cylinder V going toward the sphere of radius 2 and the -1 end going toward the origin. The atlas $\{(V,\psi)\}$ is a basis for the natural product differential structure on the patch Int $B^1 \times S^1$.

Now $(i(U),\varphi i^{-1})$ and $(j(V),\psi j^{-1})$ are charts in the structure of $X(S,e)$. Since we are interested in the orientability of $X(S,e)$, we should decide whether $\varphi i^{-1}(\psi j^{-1})^{-1}$ is orientation preserving or reversing. By a careful choice of (U,φ), we will obtain an orientation of $X(S,e)$ on the one hand, and on the other, show that $X(S,e)$ is not orientable by the criterion of Theorem 6.3. In effect, by studying the jacobian determinant of $\varphi i^{-1}(\psi j^{-1})^{-1}$, we will be able to prove both parts of the theorem.

Now, $\varphi i^{-1}(\psi j^{-1})^{-1} = \varphi i^{-1} j \psi^{-1} = \varphi e \alpha^{-1} \psi^{-1}$, so that

$$\varphi i^{-1}(\psi j^{-1})^{-1}((1+ru)v) = \varphi e(u,rv)$$

Thus to determine the orientation quality of $\varphi i^{-1}(\psi j^{-1})^{-1}$, we can look at the qualities of the two functions

$$(1+ru)v \longmapsto (u,rv)$$

and φe, which, composed, give us the coordinate transformation $\varphi i^{-1}(\psi j^{-1})^{-1}$. The quality of φe depends upon the choice of chart (U,φ) and the orientability of the surgery performed along e. However, we can explicitly determine the quality of the former. For this, note that if $(x,y) \in \psi(V)$, with $x^2 + y^2 \neq 1$, we can write (x,y) in the form $(1+ru)v$, for $u \in S^0$, $v \in S^1$, $r \in (0,1)$, where

$$u = \frac{\sqrt{x^2+y^2}-1}{|\sqrt{x^2+y^2}-1|} \qquad v = \frac{(x,y)}{\sqrt{x^2+y^2}} \qquad r = |\sqrt{x^2+y^2}-1|$$

Thus, in cartesian coordinates, the map

$$(1+ru)v \longmapsto (u,rv)$$

becomes

$$(x,y) \mapsto \left(\frac{\sqrt{x^2+y^2}-1}{|\sqrt{x^2+y^2}-1|}, \, |\sqrt{x^2+y^2}-1| \frac{(x,y)}{\sqrt{x^2+y^2}} \right)$$

Of course, the first coordinate is ±1, so the orientation quality of this map is determined by that of the map

$$(x,y) \mapsto |\sqrt{x^2+y^2}-1| \frac{(x,y)}{\sqrt{x^2+y^2}}$$

$$= \left| 1 - \frac{1}{\sqrt{x^2+y^2}} \right| (x,y)$$

whose jacobian determinant we now calculate. The jacobian matrix of this embedding is

$$\pm \begin{pmatrix} 1 - \dfrac{y^2}{(x^2+y^2)^{3/2}} & \dfrac{xy}{(x^2+y^2)^{3/2}} \\ \dfrac{xy}{(x^2+y^2)^{3/2}} & 1 - \dfrac{x^2}{(x^2+y^2)^{3/2}} \end{pmatrix}$$

where the + sign in front of the matrix is taken when $x^2+y^2 > 1$ and the - sign when $x^2+y^2 < 1$. In either case, the jacobian determinant is $1 - 1/\sqrt{x^2+y^2}$, which is positive on $\{(x,y) \mid 1 < \sqrt{x^2+y^2} < 2\}$, and negative on $\{(x,y) \mid 0 < \sqrt{x^2+y^2} < 1\}$. Thus the embedding

$$(1+ru)v \mapsto (u, rv)$$

is orientation preserving when $u = 1$ and reversing when $u = -1$.

We are now in a position to complete the proof.

Suppose, on the one hand, that e determines an orientable surgery. By definition, we may choose an orientation A of S so that $e \mid \{1\} \times \text{Int } B^2$ is orientation preserving and $e \mid \{-1\} \times \text{Int } B^2$ is orientation reversing. Let

13. Surgery on a Surface

$$B = \{(i(U),\varphi i^{-1}) \mid (U,\varphi) \in A \text{ and } U \cap e(S^0 \times \{0\}) = \phi\} \cup \{(j(V),\psi j^{-1})\}$$

It is claimed that B is a basis for an orientation of $X(S,e)$. In fact, as noted above, the coordinate transformation $\varphi i^{-1}(\psi j^{-1})^{-1}$ is the composition of

$$(1 + ru)v \mapsto (u,rv)$$

and φe. As already noted, the first is orientation preserving if $u = 1$ and reversing if $u = -1$. The second is also preserving if $u = 1$ and reversing if $u = -1$. Thus in either case the composition is orientation preserving. Clearly B is a basis, so $X(S,e)$ is orientable.

Conversely, suppose that e determines a nonorientable surgery. Choose any chart (U,φ) on S so that U is connected and meets both $e(\{1\} \times \text{Int } B^2)$ and $e(\{-1\} \times \text{Int } B^2)$ (cf. Exercise 5.6). Then by the above calculations, since $\Delta(\varphi e)$ maintains the same sign, $\Delta(\varphi i^{-1}(\psi j^{-1})^{-1})$ must change sign. Thus by Theorem 6.3, $X(S,e)$ is not orientable. □

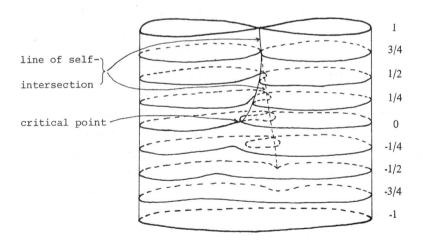

FIGURE 67

EXERCISES

1. Reconcile the two pictures of the trace of the twisting surgery of type (1,1) depicted in Figs. 61 and 62. As noted in the remark preceding Fig. 62, the critical levels appear to be different. [Hint: Consider the patch $P_{1,1}$ in Fig. 53, which is used to construct the trace. Within $P_{1,1}$, the levels change as follows: subcritical levels from the first quadrant are rectangular hyperbolas going into the fourth quadrant; the critical level from the first quadrant is a straight line into the third quadrant (with another line perpendicular to it); the supercritical levels are rectangular hyperbolas going into the second quadrant. The same features should carry over to the trace. It does in the case of Fig. 62, but apparently not in Fig. 61. The problem is that the line of self-intersection in Fig. 61 begins at the critical point. The combining of these two features in one place obscures what happens there. If these features are separated as in Fig. 67, then we may achieve the desired reconciliation.]

2. Let $e, f : \mathbb{R}^m \to \mathbb{R}^m$ be two orientation preserving embeddings. Prove that there is a diffeomorphism $h : \mathbb{R}^m \to \mathbb{R}^m$ and $\varepsilon > 0$ so that $he \mid \varepsilon B^m = f \mid \varepsilon B^m$ and h is the identity outside some compact subset of $\mathbb{R}^m$. Note that h must also be orientation preserving. [Hint: Progressively simplify to the following cases: e = identity; $f(0) = 0$; $Df(0)$ = identity.]

3. Let S be a nonorientable surface and let $e, f :$ Int $B^2 \to S$ be two embeddings. Prove that there is a diffeomorphism $h : S \to S$ and $\varepsilon > 0$ so that $he \mid \varepsilon B^2 = f \mid \varepsilon B^2$. [Hint: By improving Exercise 5.6, find a chart (U, φ) on S with $\varphi(U) = \mathbb{R}^2$, $e(3/4\, B^2) \cup f(3/4\, B^2) \subset U$, and φe and φf both orientation preserving. Apply Exercise 2 to φe and φf and transfer the diffeomorphism h back to S via φ^{-1}.]

4. Let S be a nonorientable surface and let $e, f : S^0 \times$ Int $B^2 \to S$ be two embeddings. Prove that there is a diffeomorphism $h : S \to S$ and $\varepsilon > 0$ so that $he \mid S^0 \times \varepsilon B^2 = f \mid S^0 \times \varepsilon B^2$.

5. Let S be a nonorientable surface and let $e, f : S^0 \times \text{Int } B^2 \to S$ be two embeddings. Prove that $\chi(S,e)$ is diffeomorphic to $\chi(S,f)$ [Hint: Extend the diffeomorphism h of Exercise 4 over $\chi(S,e)$.]

14
CLASSIFICATION OF ORIENTABLE SURFACES

Fig. 68 shows some orientable surfaces. The aim of this chapter is to give a systematic classification of orientable surfaces. In fact, if we continue the chain suggested by the first four pictures in Fig. 68, we obtain an infinite sequence of orientable surfaces. The classification theorem tells us that these surfaces are all distinct and that (up to homeomorphism) there are no other orientable surfaces. (That they are distinct might seem "obvious", but one might reasonably ask whether it is obvious that there is no diffeomorphism between the 99999th and 33550336th members of the sequence.)

The proof begins by showing that every orientable surface has a standard form. We start with a Morse function on the surface and alter it so that there is only one critical point of index 0 and one of index 2, with a number of index 1. The index 1 critical points are arranged so that all of the disconnecting surgeries of type (1,1) precede all of the connecting surgeries of type (1,1). Note that if height up the page represents the Morse function, then the first four surfaces in Fig. 68 almost satisfy these conditions, but the last does not. During this part of the construction, the reader should follow what is being done to the last picture, thereby seeing where it fits in our sequence.

Having standardized our surfaces, we then show that they are all distinct. This is where the surgery of type (1,2), studied in Chap. 13 is used. Basically, any one member of the sequence is obtained from its predecessor by an orientable surgery of type (1,2), so they must be distinct.

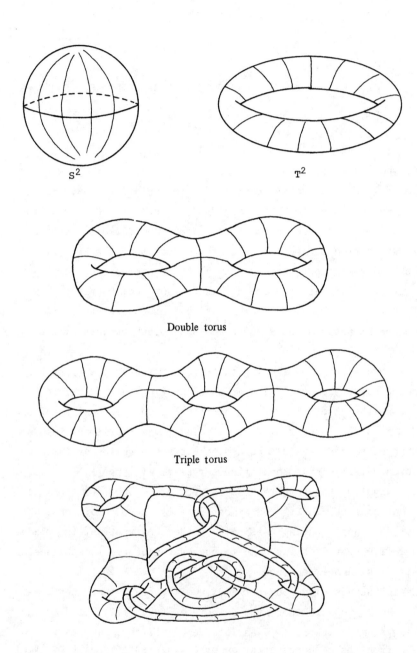

FIGURE 68

14. *Classification of Orientable Surfaces*   191

The following terminology will be useful.

Definition Let $f : S \to \mathbb{R}$ be a Morse function on a surface and let C be a component of a level of f. A *surgical descendant* of C is a component D of a higher level of f with the property that some gradientlike vector field for f has an integral curve meeting both C and D. The point of the terminology is the following. If C lies in a regular level and we perform a (1,1) surgery on C (plus another circle if the surgery is connecting) then there results either one circle or two. Either is a surgical descendant of C in much the same way as children are descendants of their parents. Furthering the analogy, any surgical descendant of a surgical descendant of C is itself a surgical descendant of C. Fig. 69 shows some surgical descendants.

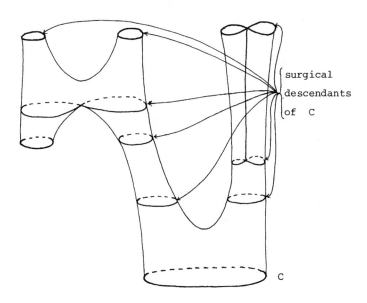

FIGURE 69

LEMMA 1. On an orientable surface there is a Morse function with one critical point of index 0, one critical point of index 2, and some critical points of index 1.

Proof: Let S be an orientable surface, and let $f : S \to \mathbb{R}$ be a Morse function with no two critical points having the same value (cf. Exercise 9.5). By Theorem 3.3 and the Heine-Borel theorem (3.2), f must have an attained minimum and an attained maximum; i.e., f has at least one critical point of each of the indices 0 and 2.

Suppose there is more than one minimum; say there are n of them. We will show how to reduce n by 1, thereby creating an inductive procedure to reduce the number to 1. A dual treatment reduces the number of maxima to 1.

Pick one of the minima, say p, and let C be the circle which results from performing the (0,2) surgery at p. Suppose there is no surgery of type (1,1) involving both C or a surgical descendant of C and the surgical descendants of all of the other n - 1 minima. The traces of all of the surgeries involving C and its descendants together with the (0,2) surgery at p forms a compact, hence closed, subset of S. So also do the traces of all of the other surgeries. Together these two closed, disjoint sets make up S, so by Corollary 2.5, S is disconnected, a contradiction. Thus some surgical descendant of C (or C itself) is connected by a (1,1) surgery to a surgical descendant of one of the other minima. Call the other minimum q and the circle which results from performing the (0,2) surgery at q, D. Let r be the critical point of index 1, the type (1,1) surgery at which first unites surgical descendants of C and D.

Let ξ be a gradientlike vector field for f, as given by Theorem 10.1. Since r is of index 1, there are two integral curves of ξ which terminate at r, one from the surgical descendants of C and one from those of D. We first show that if ξ were chosen carefully, then these integral curves would have emanated from p and q, respectively. It is enough to show how to do this for the

14. Classification of Orientable Surfaces

curve from the surgical descendants of C : call the curve γ. The curve γ must emanate from some critical point. This is a consequence of the proof of Theorem 10.2. If this is not the point p, then it must be some higher critical point, say s, which must have index 1. Take a regular level L of f above s with the property that there is no intervening critical level. Note that half of the upper sphere of s in L lies on γ. As in the proof of Theorem 10.2, the level L has a neighborhood diffeomorphic to $L \times [-1,1]$, so that the curves $\{x\} \times [-1,1]$, $x \in L$, are the integral curves of ξ within this neighborhood. Choose a point $x \in L$ which does not lie on the upper sphere of s in L. Then the integral curve of ξ through x starts below s. The point x might be quite close to the point where γ crosses L. Now modify ξ within $L \times [-1,1]$ to a gradientlike vector field ξ', so that $\xi' = \xi$ outside $L \times (-1,1)$ and the integral curve through $(x,-1)$ meets γ at $L \times \{1\}$. Fig. 70 illustrates this change—which will require an application of Lemma 4.1 to ensure smoothness of ξ'. In effect, we twist the integral curves within $L \times [-1,1]$. Now the integral curve γ' (of ξ') terminating at r emanates from a critical point below s. Repeating this process as often as necessary gives us the desired situation.

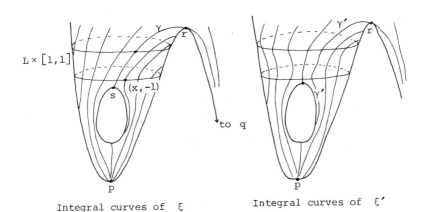

Integral curves of ξ Integral curves of ξ'

FIGURE 70

Now we have a Morse function $f : S \to \mathbb{R}$ with two minima p and q, corresponding circles C and D, and a gradientlike vector field ξ for f, such that the integral curves γ and δ terminating at the critical point r, which first unites the surgical descendants of C and D, emanate from p and q, respectively. In two steps, we modify f in a neighborhood of the curves γ and δ so that one of the minima is canceled by the saddle.

By Prototype Lemma 4.6, we can modify f to a new Morse function $f' : S \to \mathbb{R}$ so that

(a) $f' = f$ except in a neighborhood of the curves γ and δ.
(b) p and q are minima and r a (connecting) saddle of f' and no new critical points are introduced.
(c) $f'(p) = f'(q)$.
(d) There is no other critical point of f' at the same level as r.
(e) There is no other critical level between those of p and r.

This modification is achieved by decreasing the values of f within small neighborhoods of γ and δ, which will be just like the set $\alpha B^\lambda \times 2B^{m-\lambda}$ considered in Prototype Lemma 4.6. In our present case, however, we decrease, rather than increase, the values of our function f. Fig. 71 illustrates the consequence of this modification: first lower q, then lower r.

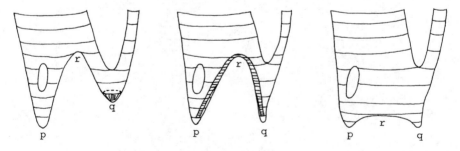

The modifications take place within the shaded regions

FIGURE 71

14. *Classification of Orientable Surfaces* 195

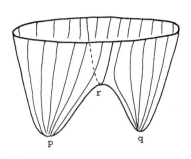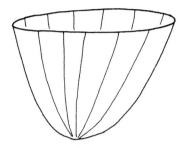

FIGURE 72

Note that the surgery determined by f' at r connects the circles C and D rather than their later surgical descendants. We complete our modifications of f by noting that the traces of two (0,2) surgeries connected by a (1,1) surgery is naturally diffeomorphic to the trace of a (0,2) surgery as in Fig. 72. We can modify f' within the traces of the surgeries at p , q , and r to f" having a single critical point (of index 0) in this region of S .

Having eliminated one minimum, we can now similarly eliminate all others (except one) and all of the maxima except one. □

LEMMA 2. Let f : S → ℝ be a Morse function on an orientable surface S such that f has only one maximum and one minimum. Then there is another Morse funcion g : S → ℝ with the same critical points as f , with the same indices, but all disconnecting surgeries determined by g precede all connecting surgeries (i.e., have a lower value).

Proof: Modify f according to Exercise 9.5 so that no two critical points of f have the same value. This does not necessitate a change of critical points nor their indices.

Suppose that some disconnecting surgery succeeds some connecting surgery. Then there is a first such disconnecting surgery: let p be the corresponding critical point. Let q be the critical point immediately below p ; then q determines a connecting surgery. Two cases arise.

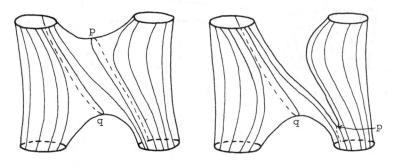

FIGURE 73

Case I. The surgery determined by p does not involve the surgical descendants of q . We can use Lemma 4.6 to modify f in a neighborhood of p so that the value at p is now less than the value at q ; cf. the similar procedure in the proof of Lemma 1.

Case II. The surgery determined by p involves the surgical descendants of q . Choose a gradientlike vector field ξ for f . If the integral curves of ξ terminating at p both come from the same one of the two circles connected by the surgery at q , then we can again use Lemma 4.6 to modify f in a neighborhood of these integral curves so that the value at p is less than the value at q ; cf. Fig. 73.

If, on the other hand, the surgery at q connects the two circles C_1 and C_2 , and one of the integral curves of ξ terminating at p comes from C_1 and the other from C_2 , then we must first modify ξ to achieve the situation depicted in Fig. 73, and then proceed as above. The modification of ξ is much the same as the modification of the gradientlike vector field carried out in the proof of Lemma 1, and we content ourselves with a picture in this situation; see Fig. 74. Note that one of the integral curves terminating at p needs to be twisted while the other is left alone. □

It is worth noting that the modification carried out in Case I of Lemma 2 may still be carried out in the situation illustrated on

14. *Classification of Orientable Surfaces*

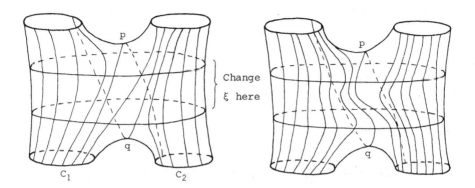

FIGURE 74

the left of Fig. 74 since, as illustated in Fig. 71, the modification merely alters the Morse function in a small neighborhood of the integral curves terminating at p . Of course the picture would not be as simple as on the right of Fig. 73 since the surface would have to pass through itself. This, however, is a shortcoming of our picture and not of the procedure for modifying the Morse function. Certainly an inadequate picture does not invalidate a procedure. We could overcome this inadequacy by resorting to the use of contour lines to represent a Morse function instead of the third coordinate, much as the Möbius band of Fig. 62 overcomes the self-intersection problem of Fig. 61. Thus on the left of Fig. 75 we have a picture of the trace of a connecting type (1,1) surgery together with a few contour lines, obtained by broadening the waist of the trousers in our previous pictures of this trace and then flattening the result when we view it from above. Changing the signs of the contour heights gives us a picture of the trace of a disconnecting surgery of type (1,1). Since these views of these surgery traces do not really allow them to be combined readily as in Figs. 73 and 74, we have an alternative view of the disconnecting surgery of type (1,1) on the right of Fig. 75. This view is effectively obtained by broadening one of the legs of the inverted trousers sufficiently to allow a complete view of the trace from above. Notice that by shrinking the connecting trace on the left, we can place it in the appropriate part of the disconnecting trace to obtain an alternative to the first picture in Fig. 73. This is the third picture of Fig. 75.

Using the third picture of Fig. 75, we can obtain another way of visualizing the procedure illustrated by Fig. 73. If the two integral curves terminating at p come from the same low level circle, then the effect of pushing p in a neighborhood of these curves below the level of q is to stretch the contours in this neighborhood so that the new height of p is correctly indicated by the stretched contours, as in Fig. 76. In Fig. 76, we see that we have set the height of p at $-\frac{1}{2}$ and the surgery at p disconnects one of the two circles making up a level below $-\frac{1}{2}$, viz., the one on the left which, at level $-\frac{1}{2}$, has become a rather stretched figure eight. Just below the level of q, a level consists of three circles, one around the inner high point, this one not being involved in the surgery at q, a second one which is connected to the third in the other original depression.

If the two integral curves terminating at p come from different circles below the level of q, then we can still lower p in a neighborhood of these curves until it is below the level of q by stretching the contours as above. This time, however, the surgery at p will be a connecting surgery! By lowering the critical point p we have changed the kind of surgery. One can check that although we have done nothing to the Morse function near q, we have changed that surgery from a connecting surgery to a disconnecting surgery. Thus by repeating the process which worked in the previous situation we have gained nothing.

Lemma 2 above has now put our Morse functions, and hence surgeries, into a standard form. In order to complete our classification, two points are necessary. We must show that two surfaces formed by performing a surgery of type $(0,2)$, k disconnecting surgeries of type $(1,1)$, k connecting surgeries of type $(1,1)$, and then a surgery of type $(2,0)$ are the same; we must show that two surfaces formed as above, but with the second involving only ℓ

14. Classification of Orientable Surfaces 199

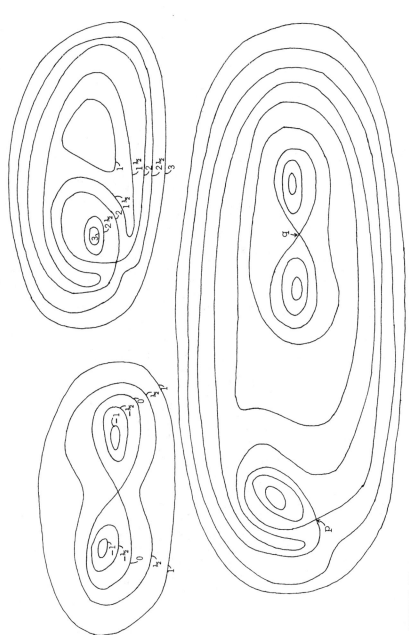

FIGURE 75

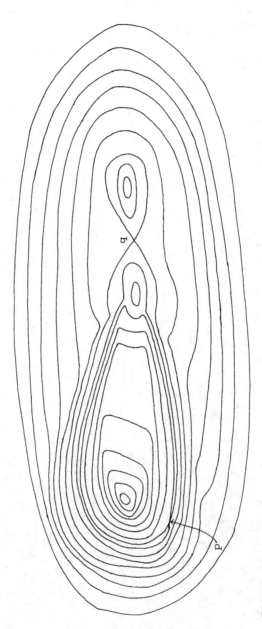

FIGURE 76

14. Classification of Orientable Surfaces

connecting and disconnecting type (1,1) surgeries ($\ell < k$), are different. Our completion of these two tasks involves the orientable type (1,2) surgeries considered in the last chapter.

Definition A *sphere with n handles* is a surface obtained from S^2 by performing n orientable type (1,2) surgeries. If S is a sphere with handles then the performance of an orientable type (1,2) surgery on S will be called *adding a handle* to S. Note that, by Lemma 13.5, a sphere with handles is orientable. Figs 51 and 65 illustrate why the process is called adding a handle.

We will show that a surface obtained by performing one surgery of each of types (0,2) and (2.0) and k surgeries of each of the disconnecting and connecting type (1,1) is a sphere with k handles. The two points noted above will then be satisfied in the context of spheres with handles.

PROPOSITION 3. Every orientable surface is homeomorphic to a sphere with handles.

Proof: Let S be an orientable surface and let $f : S \to \mathbb{R}$ be a Morse function given by Lemma 2. Suppose that f has 2k saddle points (hence k of each of the disconnecting and connecting type (1,1) surgeries). We prove by induction on k that S is a sphere with k handles, the case $k = 0$ following from Exercise 10.7.

Now assume $k \geq 1$. Let L be a regular level above the highest disconnecting saddle and below the lowest connecting saddle. Then L consists of $k + 1$ circles. Let C be one of these circles (see Fig. 77). As in Theorem 10.2, C has a neighborhood diffeomorphic to $C \times [-1,1]$, with $\{x\} \times [-1,1]$ being part of an integral curve, $x \in C$. Apply surgery of type (2,1) to S around C to get $\tilde{S}$, where the 1-sphere C is replaced by the 0-sphere $P \subset \tilde{S}$.

Following the ideas of Lemma 1, we can cancel the newly introduced minimum (one point of P) with one of the disconnecting saddles and the newly introduced maximum with one of the connecting saddles to obtain on S a Morse function with one minimum, one maximum, and $2(k - 1)$ saddles. Thus by the inductive hypothesis, S is a sphere with $k - 1$ handles.

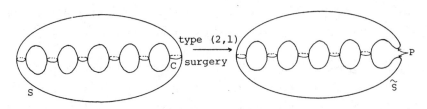

FIGURE 77

Reversing the (2,1) surgery from S to $\tilde{S}$, we may perform an orientable (1,2) surgery to get from $\tilde{S}$ to S. Thus S is a sphere with k handles. □

Remark If we could have replaced "homeomorphic" by "diffeomorphic" in the case k = 0 above, i.e., in Exercise 10.7, then we could have made a similar change in the statement of Proposition 3. This change can be made, but not here.

PROPOSITION 4. Any two spheres with n handles are diffeomorphic.

Proof: Use induction on n. When n = 0, the result is trivial.

Let S be a sphere with n - 1 handles. It suffices to show that all manifolds obtained from S by adding a handle are diffeomorphic. Let $e, e' : S^0 \times \text{Int } B^2 \to S$ be two embeddings such that the type (1,2) surgeries performed along them are both orientable. We might as well assume that the orientations are such that e and e' both preserve orientation on $\{1\} \times \text{Int } B^2$ and reverse orientation on $\{-1\} \times \text{Int } B^2$. Let $f : S \to S$ be a diffeomorphism such that e and fe' agree on $S^0 \times \varepsilon B^2$ (see Exercise 2). Then the manifolds $\chi(S,e)$ and $\chi(S,fe')$ are almost the same. Certainly they are naturally diffeomorphic. Moreover, f determines a natural diffeomorphism from $\chi(S,e')$ to $\chi(S,fe')$. Thus $\chi(S,e)$ is diffeomorphic to $\chi(S,e')$. □

A converse of Proposition 4 holds, and we now formulate and prove it.

14. Classification of Orientable Surfaces

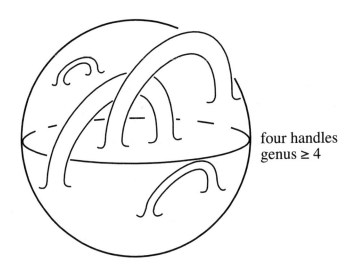

four handles
genus ≥ 4

Definition The *genus* of an orientable surface S is the maximum number of embeddings $e_i : S^1 \times (-1,1) \to S$ such that for $i \neq j$, $e_i(S^1 \times (-1,1)) \cap e_j(S^1 \times (-1,1)) = \phi$ and $S - \cup_i e_i(S^1 \times (-1,1))$ is connected.

The domain $S^1 \times (-1,1)$ in the above definition could be replaced by S^1 but we will find it more convenient to use $S^1 \times (-1,1)$. Note that genus is a topological property (of surfaces only!).

As suggested by Fig. 78, a sphere with n handles has genus $\geq n$. On the other hand, the genus of an orientable surface is finite. The Jordan curve theorem implies that S^2 has genus 0.

LEMMA 5. A sphere with n handles has genus n.

Proof: We need only prove that the genus does not exceed n, and this is done by induction on n. Since the sphere with 0 handles is S^2 which has genus 0, the induction begins.

Suppose S is a sphere with n handles (n > 0) and that spheres with m handles, $m < n$, have genus $\leq m$. If the genus of S is ℓ, then there are ℓ embeddings $e_i : S^1 \times (-1,1) \to S$ such that for

$i \neq j$, $e_i(S^1 \times (-1,1)) \cap e_j(S^1 \times (-1,1)) = \phi$ and
$S - \bigcup_{i=1}^{\ell} e_i(S^1 \times (-1,1))$ is connected. Since $\ell \geq n > 0$, there is at least one embedding e_i. Consider $\chi(S, e_\ell) = T$, say. Since T is connected, by Exercise 3, T is a sphere with m handles for some $m < n$. Thus by inductive hypothesis, T has genus $< n$. Now the embeddings e_i, $i = 1, \ldots, \ell - 1$ may also be considered as embeddings of $S^1 \times (-1,1)$ in T, with $e_i(S^1 \times (-1,1)) \cap e_j(S^1 \times (-1,1)) = \phi$ for $i \neq j$ and $T - \bigcup_{i=1}^{\ell-1} e_i(S^1 \times (-1,1))$ connected. Thus the genus of T is at least $\ell - 1$. Hence we have $\ell - 1 < n$, so that $\ell \leq n$, as required. □

COROLLARY 6. If $m \neq n$, then a sphere with m handles is not diffeomorphic to a sphere with n handles.

COROLLARY 7 (Classification of Orientable Surfaces). Every orientable surface is (homeomorphic to) a sphere with n handles for some n; moreover, no two spheres with different numbers of handles are homeomorphic.

Remark As in the remark following Proposition 3, "homeomorphic" could be replaced by "diffeomorphic."

EXERCISES

1. Prove that the modification of the vector field ξ required in the proof of Lemma 1 and illustrated in Fig. 70 can be carried out.

2. Let S be an oriented surface and let $e, f : S^0 \times \text{Int } B^2 \to S$ be two embeddings such that e and f both preserve orientation on $\{1\} \times \text{Int } B^2$ and both reverse orientation on $\{-1\} \times \text{Int } B^2$. Prove that there is a diffeomorphism $h : S \to S$ and $\varepsilon > 0$ such that $he \mid S^0 \times \varepsilon B^2 = f \mid S^0 \times \varepsilon B^2$. [Hint: Compare with Exercise 13.4].

3. Prove that if S is a sphere with n handles and T is a (connected) surface obtained by performing a surgery of type (2,1) on S, then T is a sphere with m handles for some $m < n$. [Hints: Proposition 3 assures that T is a sphere with m handles for

some m. If $m \geq n$, then by performing $m - n$ further well-chosen surgeries of type (2,1), by Proposition 4 we can return to S. Thus by performing $m + 1 - n$ surgeries of type (2,1) on S we get back to S. We can repeat this as often as we like, say ℓ times for any ℓ. This means that the genus of S is at least $\ell(m + 1 - n)$, violating finiteness.]

WHITNEY'S EMBEDDING THEOREM

Throughout this chapter we will assume that M^m is a compact submanifold of $\mathbb{R}^n$ for some n.

Our aim is to show that every compact m-manifold embeds in $\mathbb{R}^{2m+1}$. The result extends to noncompact manifolds with mild restrictions and can be improved to $\mathbb{R}^{2m}$, but we will not consider either of these two cases. One cannot in general improve the result beyond $\mathbb{R}^{2m}$; for example, S^1 does not embed in $\mathbb{R}^1$ and nonorientable surfaces do not embed in $\mathbb{R}^3$. However there are large classes of manifolds for which there is some improvement, for example our classification of orientable surfaces in Chap. 14 implies that they all embed in $\mathbb{R}^3$, and S^m embeds in $\mathbb{R}^{m+1}$.

There are two main stages in the proof of Whitney's embedding theorem, the first of which was carried out in Chap. 7. Theorem 7.2. tells us that every compact m-manifold embeds in $\mathbb{R}^n$ for sufficiently large n. This justifies our assumption above that M is a submanifold of $\mathbb{R}^n$ for some n. The second stage in the proof involves a reduction of the number n to 2m + 1 . This proceeds inductively. Provided n > 2m + 1 , we show how to embed M in $\mathbb{R}^{n-1}$.

Here is an outline of the ideas behind the second stage of the proof. For $u \in S^{n-1} \subset \mathbb{R}^n$, let

$$H_u = \{x \in \mathbb{R}^n \mid x \cdot u = 0\}$$

The · here denotes the usual scalar product, so that H_u is the hyperplane orthogonal to the vector u . Thus H_u is diffeomorphic (by a linear diffeomorphism) to $\mathbb{R}^{n-1}$. Define $\pi_u : \mathbb{R}^n \to H_u$ by $\pi_u(y) = y - (y \cdot u)u$. Thus $\pi_u(y)$ is the point of H_u nearest to y. Imagine a sun shining in $\mathbb{R}^n$ so that the rays are parallel to the line from 0 to u. Then $\pi_u(y)$ is the shadow of y on H_u. We will show that there is some $u \in S^{n-1}$ for which $\pi_u | M$ is an embedding, i.e., the shadow $\pi_u(M)$ looks just like M, provided that $n > 2m + 1$. Since H_u is diffeomorphic to $\mathbb{R}^{n-1}$, we will be through. In fact almost any choice of $u \in S^{n-1}$ will do. As noted in Chap. 5, using Theorem 3.6, we need to show that $\pi_u | M$ is an injective immersion. We will show that the two sets

$$\{u \in S^{n-1} \mid \pi_u | M \text{ is not injective}\}$$

$$\{u \in S^{n-1} \mid \pi_u | M \text{ is not an immersion}\}$$

are both small in a sense to be described below, provided $n > 2m + 1$. From this the result will follow.

Actually we will show that $\pi_u | M$ is an immersion for most $u \in S^{n-1}$ even when $n > 2m$, so that the procedure will give us an immersion of M in $\mathbb{R}^{2m}$. Unfortunately we do not obtain an injection of M in $\mathbb{R}^{2m}$ using this procedure. Fig. 79 gives (a projection on $\mathbb{R}^2$ of) an embedding of S^1 in $\mathbb{R}^3$. For each $u \in S^2$, π_u is not injective on this embedded circle, so for this embedding π_u could not possibly give rise to an embedding of S^1 in $\mathbb{R}^2$.

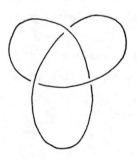

FIGURE 79

15. Whitney's Embedding Theorem

The "smallness" mentioned above is that of Hausdorff measure 0, which we now introduce.

Definition A subset S of $\mathbb{R}^p$ has *q-dimensional Hausdorff measure* 0 iff for all $\varepsilon > 0$, there is a sequence $\{B(x_i; r_i) \mid i = 1, 2, \ldots\}$ of open balls in $\mathbb{R}^p$ covering S so that $\sum_{i=1}^{\infty} r_i^q < \varepsilon$. The term "q-dimensional" is used because, up to a constant multiple, r^q is the q-dimensional volume of the ball of radius r (for q = 1, 2, 3, 4, the constant is, respectively, 2, π, $4\pi/3$, $\pi^2/2$). Since the constant could be absorbed in the ε, we may leave it out. The definition tells us that a set has q-dimensional Hausdorff measure 0 provided we can cover it by balls whose q-dimensional volumes can be made arbitrarily small. If S has q-dimensional Hausdorff measure 0, we will write $\theta_q(S) = 0$.

Examples It is false that $\theta_1(S^1) = 0$, since the circumference of S^1 is 2π, so that if $\{B(x_i; r_i) \mid i = 1, 2, \ldots\}$ is a sequence of open balls in $\mathbb{R}^2$ covering S^1, then $\sum_{i=1}^{\infty} r_i > \pi$. For $q > 1$, $\theta_q(S^1) = 0$; for example, the area of S^1 is zero. In fact, if S is any subset of $\mathbb{R}^p$ then $\theta_q(S) = 0$ for $q > p$, since $\theta_q(\mathbb{R}^p) = 0$ for $q > p$, and it is obvious that any subset of a set of q-dimensional Hausdorff measure 0 also has q-dimensional Hausdorff measure 0.

The following result is the only one in this text which shows a striking difference between smooth functions and continuous functions.

LEMMA 1. Let $f : U \to \mathbb{R}^q$ be a smooth function with U an open subset of $\mathbb{R}^q$ and let $S \subset U$. If $\theta_q(S) = 0$, then $\theta_q(f(S)) = 0$.

Proof: *Case I*. Suppose that there is a compact convex subset K of U with $S \subset K$. ("Convex" means that if $x, y \in K$ then the whole of the line segment from x to y lies in K.) Since K is compact, by the Heine-Borel theorem, there exists $b \in \mathbb{R}$ such that for all i, j, and all $x \in K$,

$$\left|\frac{\partial f_i}{\partial x_j}\bigg|_x\right| \leq b$$

By Exercise 1,

(*) $\quad |f(x) - f(y)| \leq bq\,|x - y| \qquad$ for all $x, y \in K$

Let $\varepsilon > 0$ be given. Since $\theta_q(S) = 0$, there is a sequence $\{B(x_i; r_i) \mid i = 1, 2, \ldots\}$ of open balls in $\mathbb{R}^q$ so that $S \subset \bigcup_i B(x_i; r_i)$ and

$$\sum_{i=1}^{\infty} r_i^q < \frac{\varepsilon}{(bq)^q}$$

By (*), we have

$$f(B(x;r) \cap K) \subset B(f(x); bqr) \qquad \text{for all } x \in K$$

so

$$f(S) \subset \bigcup_{i=1}^{\infty} f(B(x_i; r_i) \cap K)$$
$$\subset \bigcup_{i=1}^{\infty} B(f(x_i); bqr_i)$$

But, $\sum_{i=1}^{\infty} (bqr_i)^q = (bq)^q \sum_i r_i^q < \varepsilon$. Hence $\theta_q(f(S)) = 0$ in this case.

Case II. (General case). This requires two observations: a countable union of sets of q-dimensional Hausdorff measure 0 also has q-dimensional Hausdorff measure 0; any open subset of $\mathbb{R}^p$ is a countable union of compact convex subsets. The proof of the first observation is straightforward, and some hints for the proof of the second are given in Exercise 3. The result follows from case I and these two observations. □

COROLLARY 2. Let $f : U \rightarrow \mathbb{R}^q$ be a smooth function, where U is an open subset of $\mathbb{R}^p$, $p < q$. Then $\theta_q(f(U)) = 0$.

15. Whitney's Embedding Theorem

Proof: Let $\hat{U} = \{(x,y) \in \mathbb{R}^q \mid x \in U \text{ and } y \in \mathbb{R}^{q-p}\}$. Then $\hat{U}$ is an open subset of $\mathbb{R}^q$, and U may be thought of as a subset of $\hat{U}$: let the second coordinate be 0. Note that $\theta_q(U) = 0$. Define $\hat{f} : \hat{U} \to \mathbb{R}^q$ by $\hat{f}(x,y) = f(x)$. Then by Lemma 1, $\theta_q(\hat{f}(U)) = 0$, i.e., $\theta_q(f(U)) = 0$. □

Remark Lemma 1 and Corollary 2 are false if we require f merely to be continuous, since there are space-filling curves. Let $f : (-1,1) \to \text{Int } B^2$ be a Peano curve, i.e., a continuous surjection. Then $\theta_2(f(-1,1)) = 0$ is false. The proof breaks down at (*), this inequality being false in general for continuous functions.

The concept of Hausdorff measure 0 easily transfers to manifolds via charts.

Definition Let $(N^n, \mathcal{D})$ be a manifold and $S \subset N$. Then S has q-dimensional Hausdorff measure 0, $\theta_q(S) = 0$, iff for every chart $(U, \varphi) \in \mathcal{D}$, $\theta_q(\varphi(S \cap U)) = 0$.

By Lemma 1, we could replace $\mathcal{D}$ by any basis for $\mathcal{D}$. The following is the key lemma in our proof of Whitney's embedding theorem.

LEMMA 3. Let $(N^n, \mathcal{D})$ and (P^p, E) be two differentiable manifolds with $n < p$ and N compact. Let U be an open subset of N and $f : U \to P$ a smooth function. Then $\theta_p(f(U)) = 0$.

Proof: Let $(V, \psi) \in E$. We must show that $\theta_p(\psi(f(U) \cap V) = 0$. Since N is compact, there are finitely many charts $(U_i, \varphi_i) \in \mathcal{D}$, $i = 1, \ldots, \ell$, so that $N = \bigcup_{i=1}^{\ell} U_i$. Thus

$$f(U) = \bigcup_{i=1}^{\ell} f(U \cap U_i)$$

so

$$\psi(f(U) \cap V) = \bigcup_{i=1}^{\ell} \psi(f(U \cap U_i) \cap V)$$

Consider $\psi f \varphi_i^{-1} : \varphi_i(U \cap U_i \cap f^{-1}(V)) \to \mathbb{R}^p$. Since U, U_i, and $f^{-1}(V)$ are open subsets of N, the domain of $\psi f \varphi_i^{-1}$ is an open subset of $\mathbb{R}^n$. Furthermore $\psi f \varphi_i^{-1}$ is smooth and $n < p$, so by Corollary 2,

$$\theta_p(\psi f \varphi_i^{-1}(\varphi_i(U \cap U_i \cap f^{-1}(V)))) = 0$$

i.e.,

$$\theta_p(\psi(f(U \cap U_i) \cap V)) = 0$$

Thus $\theta_p(\psi(f(U) \cap V)) = 0$ as required. □

Recall our assumption that M^m is a compact submanifold of $\mathbb{R}^n$. Recall also π_u.

LEMMA 4. If $n > 2m + 1$, then

$$\theta_{n-1}(\{u \in S^{n-1} \mid \pi_u | M \text{ is not injective}\}) = 0$$

Proof: Suppose $x, y \in \mathbb{R}^n$ with $x \neq y$. Then $\pi_u(x) = \pi_u(y)$ iff $x - y$ is a multiple of u iff $(x - y)/|x - y| = \pm u$. Thus if $\pi_u | M$ is not injective, then there are $x, y \in M$ with $x \neq y$ such that $(x - y)/|x - y| = u$.

If we define $F : M \times M - \{(x,x) \mid x \in M\} \to S^{n-1}$ by $F(x,y) = (x - y)/|x - y|$, then we have just shown that if $u \in S^{n-1}$ is such that $\pi_u | M$ is not injective, then $u \in F(M \times M - \{(x,x) \mid x \in M\})$ Now $M \times M$ is a $2m$-manifold which we shall show presently to be compact. Since the domain of F is an open subset of $M \times M$ and $2m < n - 1$, Lemma 3 implies that

$$\theta_{n-1}(F(M \times M - \{(x,x) \mid x \in M\})) = 0$$

from which the result follows.

The compactness of $M \times M$ follows from a general topological theorem which says that the product of two compact topological spaces is compact. This result may be proved by modifying the proof given in Appendix A that the unit cube C^n is compact. Alternatively we

15. Whitney's Embedding Theorem

can deduce the compactness of M × M from the Heine-Borel theorem as follows. By the Heine-Borel theorem, M is a bounded subset of $\mathbb{R}^n$, so M × M is a bounded subset of $\mathbb{R}^n \times \mathbb{R}^n = \mathbb{R}^{2n}$. Further, M is a closed subset of $\mathbb{R}^n$, so

$$\mathbb{R}^{2n} - M \times M = [(\mathbb{R}^n - M) \times \mathbb{R}^n] \cup [\mathbb{R}^n \times (\mathbb{R}^n - M)]$$

is a union of open subsets of $\mathbb{R}^{2n}$, hence M × M is a closed subset of $\mathbb{R}^{2n}$. Thus by the Heine-Borel theorem, M × M is compact. □

We now turn to the immersivity part of the proof. Exercise 8.4 motivates us to use the tangent manifold. Recall that the tangent manifold TM was given a differential structure in Exercise 10.1 which made it into a 2m-manifold. The embedding of M in $\mathbb{R}^n$ gives us an embedding $E : TM \to \mathbb{R}^{2n} = \mathbb{R}^n \times \mathbb{R}^n$ as follows. Given $(p,v) \in TM$, Proposition 8.2 gives a natural embedding $e_p : TM_p \to \mathbb{R}^n$, where $e_p(TM_p)$ is the m-dimensional affine subspace of $\mathbb{R}^n$ containing p and all of the tangent lines at p of curves in M passing through p. Let $E(p,v) = (p, e_p(v))$.

Define $\hat{E} : TM \to \mathbb{R}$ as follows: $\hat{E}(p,v)$ is the length of the tangent vector v; more precisely,

$$\hat{E}(p,v) = |e_p(v) - p|$$

Let $T_1 M = \{(p,v) \in TM \mid \hat{E}(p,v) = 1\}$.

The manifold TM and its subset $T_1 M$ are a bit hard to visualize even when $M = S^1$. In such a case, $S^1 \subset \mathbb{R}^2$ and our embedding E embeds TS^1 in $\mathbb{R}^4$. We can, however, embed TS^1 in $\mathbb{R}^3$ as follows: for each $p \in S^1$, rotate $e_p(TS^1_p)$ through 90° about p, so that the line is now parallel with the third coordinate axis and what was the counterclockwise direction of $e_p(TS^1_p)$ now points in the positive direction of the third coordinate. This rotation is compatible with the differential structure imposed on TS^1 in Exercise 10.1 and embeds TS^1 as a cylinder over S^1 in $\mathbb{R}^3$. Thus TS^1 is diffeomorphic to $S^1 \times \mathbb{R}$ in a natural way. Of course $S^1 \times \mathbb{R}$ is just an annulus in $\mathbb{R}^2$, so TS^1 embeds in $\mathbb{R}^2$. The natural diffeomorphism from TS^1 to $S^1 \times \mathbb{R}$ carries $T_1 S^1$ to the pair of circles $S^1 \times S^0$.

We might try a similar method of visualizing TS^2, although of course TS^2 is a 4-manifold. Even so, it is not that simple, and in fact, TS^2 is not diffeomorphic to $S^2 \times \mathbb{R}^2$, although it is locally; i.e., each $p \in S^2$ has a neighborhood U such that $\pi^{-1}(U)$ is naturally diffeomorphic to $U \times \mathbb{R}^2$. Similarly $T_1 S^2$ is not diffeomorphic to $S^2 \times S^1$, although again this is locally the case.

LEMMA 5. $T_1 M$ is a compact $(2m - 1)$-manifold.

Proof: First $T_1 M$ is a $(2m - 1)$-manifold, for 1 is a regular value of $\hat{E}$, since $\hat{E}|\{(p,tv) | t \in \mathbb{R}\}$ is of rank 1 at (p,v) for all $(p,v) \in T_1 M$. Hence by Corollary 9.2, $T_1 M$ is a $(2m - 1)$-manifold.

To prove compactness of $T_1 M$, we use the Heine-Borel theorem. Let $T_1 M_p = \pi^{-1}(p) \cap T_1 M$. Then we can define a function $\alpha : M \to \mathbb{R}$ by

$$\alpha(p) = \max\{|E(p,v)| \;\big|\; v \in T_1 M_p\}$$

The maximum exists because $T_1 M_p$ is compact. In fact it is homeomorphic to S^{m-1}. Moreover, α is continuous, so, since it has a compact domain, it is bounded, say by a; for all $p \in M$, $\alpha(p) \leq a$. Then $E(T_1 M)$ lies within a of 0, so is bounded.

Next, $E(TM)$ is closed. For suppose $(x,y) \in \mathbb{R}^n \times \mathbb{R}^n - E(TM)$. If $x \notin M$, then $(\mathbb{R}^n - M) \times \mathbb{R}^n$ is an open neighborhood of (x,y) lying in $\mathbb{R}^n \times \mathbb{R}^n - E(TM)$. If $x \in M$, then $y \notin e_x(TM_x)$. Now $e_x(TM_x)$ is a linear, hence closed, subspace of $\mathbb{R}^n$. Thus there exists $r > 0$ so that $B(y;r) \cap e_x(TM_x) = \phi$. Hence there is a neighborhood U of x in $\mathbb{R}^n$ so that for all $z \in U \cap M$ and all $v \in e_z(TM_z)$, $|y - v| > r/2$. Then the set $U \times B(y;r/2)$ is a neighborhood of (x,y) which does not meet $E(TM)$. Thus $E(TM)$ is closed.

By continuity of $\hat{E}$, $T_1 M$ is a closed subset of TM, so, since E is an embedding, $E(T_1 M)$ is closed in $E(TM)$, and hence in $\mathbb{R}^{2n}$.

Since $E(T_1 M)$ is bounded and closed in $\mathbb{R}^{2n}$, by the Heine-Borel theorem, it is compact. Finally, E is an embedding; hence $T_1 M$ is also compact. □

15. Whitney's Embedding Theorem

LEMMA 6. If $n > 2m$, then

$$\theta_{n-1}(\{u \in S^{n-1} \mid \pi_u|M \text{ is not an immersion}\}) = 0$$

Proof: Define $F : T_1M \to S^{n-1}$ by $F(p,v) = e_p(v) - p$. Since T_1M has dimension $2m - 1 < n - 1$, by Lemma 3, $\theta_{n-1}(F(T_1M)) = 0$. Thus it remains to show that

$$\{u \in S^{n-1} \mid \pi_u|M \text{ is not an immersion}\} \subset F(T_1M)$$

Suppose $u \in S^{n-1}$ and $\pi_u|M$ is not an immersion. Then there exists $x \in M$ at which $\pi_u|M$ does not have rank m. For the moment, we will drop the subscript u from π_u and H_u. Then by Exercise 8.4, $d\pi_x : TM_x \to TH_{\pi(x)}$ does not have rank m, i.e., is not a monomorphism. Thus there exists $v \in TM_x$, $v \neq 0$, with $d\pi_x(v) = 0$. By multiplying v by a scalar if necessary, we may assume that $v \in T_1M_x$. Let γ be a curve on M through $x = \gamma(t)$ such that v is the velocity vector of γ at t. Then $F(v) = \gamma'(t)$.

Since $d\pi_x(v)$ is the velocity vector of $\pi\gamma$ at t and $d\pi_x(v) = 0$, we have that $(\pi\gamma)'(t) = 0$, so, since π is linear, $\gamma'(t) = \pm u$. Thus $u = F(\pm v) \in F(T_1M)$, which completes the proof of the above set inclusion. □

THEOREM 7 (Whitney's Embedding Theorem). Let M^m be a compact manifold. Then M embeds in $\mathbb{R}^{2m+1}$.

Proof: As already noted, by Theorem 7.2, M embeds in $\mathbb{R}^n$ for some n, say $e : M \to \mathbb{R}^n$ is an embedding. If $n \leq 2m + 1$, we are through. If $n > 2m + 1$, then by Lemmas 4 and 6,

$$\theta_{n-1}(\{u \in S^{n-1} \mid \pi_u|M \text{ is not an injective immersion}\}) = 0$$

Thus by Exercise 4, there exists $u \in S^{n-1}$ with $\pi_u|M$ an injective immersion, i.e., an embedding, since M is compact. Thus π_u embeds M in H_u, which is diffeomorphic to $\mathbb{R}^{n-1}$, so M embeds in $\mathbb{R}^{n-1}$. Continuing this process, we obtain an embedding of M in $\mathbb{R}^{2m+1}$. □

Remark Lemma 4 forces the process in the above proof to stop. In particular, we could apply Lemma 6 again to obtain an immersion of M in $\mathbb{R}^{2m}$; cf. Fig. 79.

EXERCISES

1. Let U be a convex subset of $\mathbb{R}^n$ and $f : U \to \mathbb{R}^n$ a C^1 function, and suppose that there is a real number b satisfying

$$\left| \frac{\partial f_i}{\partial x_j} \right| \leq b \qquad \text{for all } i,j = 1, \ldots, n$$

throughout U. Prove that for all $x,y \in U$, $|f(x) - f(y)| \leq bn|x - y|$. [Hint: Use the mean value theorem, which says that for all $x,y \in U$ and all $i = 1, \ldots, n$, there exists z_i on the line segment from x to y (thus in U!) so that $f_i(x) - f_i(y) = Df_i(z_i)(x - y)$.]

2. Prove that a countable union of sets of n-dimensional Hausdorff measure 0 also has n-dimensional Hausdorff measure 0.

3. Prove that any open subset of $\mathbb{R}^n$ is a countable union of compact convex subsets. [Hint: Given an open subset U of $\mathbb{R}^n$, let

$$A = \{(x_1, \ldots, x_n) \in U \mid \text{for all } i, x_i \in \mathbb{Q}\}$$
$$A = \{Cl\ B(x;r) \mid x \in A, r \in \mathbb{Q}, \text{and } Cl\ B(x;r) \subset U\}$$

Prove that A is a countable collection of compact convex sets whose union is U.]

4. Prove that if S is a subset of the manifold M^m and $\theta_m(S) = 0$, then $Cl\ (M - S) = M$. In particular, unless $M = \phi$, $S \neq M$.

Appendix A

THE UNPROVED THEOREMS

In this appendix, we either prove the theorems which were not proved in the body of the text or else give outlines of the proofs together with sources in which complete proofs can be found.

THEOREM 3.2 (Heine-Borel Theorem). A subset of $\mathbb{R}^n$ is compact iff it is closed and bounded.

Proof: $\Rightarrow$: Compact subsets of the Hausdorff space $\mathbb{R}^n$ are closed by Theorem 3.5. The family $\{B(0;r) | r \in \mathbb{R}, r > 0\}$ forms an open cover of every subset of $\mathbb{R}^n$, in particular, of any compact subset. By taking a finite subcover, we see that any compact subset of $\mathbb{R}^n$ is contained in $B(0;r)$ for some $r \in \mathbb{R}$; hence compact subsets are bounded.

$\Leftarrow$: By Theorem 3.5 and Corollary 3.4, it is enough to show that the unit cube

$$C^n = \{(x_1, \ldots, x_n) \in \mathbb{R}^n \mid \text{for all } i, \ 0 \leq x_i \leq 1\}$$

is compact, since any bounded subset of $\mathbb{R}^n$ is contained in some homeomorph of C^n. Our proof proceeds by induction on n, the case $n = 1$ having been taken care of in Exercise 3.6. Let F be an open cover of C^n. Then for all $x \in C^n$, there is an open set U_x of the form

$$\{(y_1, \ldots, y_n) \mid |y_i - x_i| < \varepsilon(x), \text{ for all } i = 1, \ldots, n\}$$

for some $\varepsilon(x) > 0$ so that U_x is contained in some member of F.

Let $U = \{U_x \mid x \in C^n\}$. Then U is an open cover of C^n. It is enough to find a finite subcover of U, for this will lead to a finite subcover of F.

For each $t \in [0,1]$, let

$$C^n_t = \{(t, x_2, \ldots, x_n) \in C^n\}$$

Note that C^n_t is homeomorphic to C^{n-1} and U is an open cover of C^n_t. Thus there is a finite subcover, say U_t, of C_t (see Fig. 80). Let ε_t be the smallest $\varepsilon(x)$ corresponding to the sets U_x which form U_t. Since there are only finitely many, $\varepsilon_t > 0$. The finite family U_t actually covers $\cup \{C^n_s \mid |t - s| < \varepsilon_t\}$.

Since $\{(t - \varepsilon_t, t + \varepsilon_t) \mid t \in [0,1]\}$ is an open cover of $[0,1]$, by Exercise 3.7 it has a finite subcover, say

$$\{(t_i - \varepsilon_{t_i},\ t_i + \varepsilon_{t_i}) \mid i = 1, \ldots, k\}$$

Then $\cup_{i=1}^{k} U_{t_i}$ is a finite subfamily of U covering C^n, so C^n is compact. □

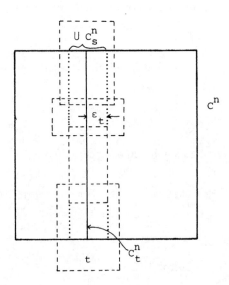

FIGURE 80

Appendix A. The Unproved Theorems

THEOREM 4.2 (Inverse Function Theorem). Let U be an open subset of $\mathbb{R}^m$ and let $f : U \to \mathbb{R}^m$ be a C^r function. Let $x_0 \in U$ and suppose that $Df(x_0)$ is nonsingular. Then f is a C^r diffeomorphism of some neighborhood of x_0 onto some neighborhood of $f(x_0)$.

Proof: We might as well assume that $Df(x_0)$ is the identity. For if not, let $\lambda : \mathbb{R}^m \to \mathbb{R}^m$ be the linear transformation with matrix representation $Df(x_0)$. Since $Df(x_0)$ is nonsingular, λ is an isomorphism, hence a C^∞ diffeomorphism. Thus if the theorem is true for $\lambda^{-1} f$, it is also true for f. Note, by the chain rule, that $D(\lambda^{-1} f)(x_0) = 1$, so it suffices to prove the result for C^r functions whose jacobian at x_0 is the identity.

Assume, then, that $Df(x_0) = 1$. There is a closed ball, say A, centered at x_0 such that
i. $Df(x)$ is nonsingular for all $x \in A$.
ii. $\left| (\partial f_i / \partial x_j) \big|_x - (\partial f_i / \partial x_j) \big|_{x_0} \right| < 1/2m^2$ for all i,j and all $x \in A$.

These claims follow from the fact that $Df(x_0)$ is nonsingular and f is C^1.

Define $g : U \to \mathbb{R}^m$ by $g(x) = f(x) - x$. Then by ii and Exercise 15.1, we have

$$|g(x) - g(y)| \leq \tfrac{1}{2} |x - y| \quad \text{for all } x,y \in A$$

so that, by the triangle inequality,

iii. $|x - y| \leq 2 |f(x) - f(y)|$ for all $x,y \in A$.

Now $f(\partial A)$ is a compact set, and by iii, $f(x_0) \notin f(\partial A)$. Thus there exists $\alpha > 0$ such that for all $x \in \partial A$, $|f(x) - f(x_0)| \geq \alpha$. Let

$$B = \{y \mid |y - f(x_0)| < \tfrac{\alpha}{2}\}$$

Then

iv. $|y - f(x_0)| < |y - f(x)|$ for all $x \in \partial A$ and all $y \in B$.

Let $C = \text{Int } A \cap f^{-1}(B)$. Then C is a neighborhood of x_0. We show that f is a C^r diffeomorphism of C onto B.

$\quad f : C \to B$ *is injective*: This follows immediately from iii.

$\quad f : C \to B$ *is surjective*: Let $y \in B$. Define $h : A \to \mathbb{R}$ by

$$h(x) = |y - f(x)|^2 = \sum_i [y_i - f_i(x)]^2$$

Since h is continuous, it attains its minimum on A. By iv, this minimum cannot be in ∂A, so it must be in Int A, where h is differentiable. Let h attain its minimum at $x \in \text{Int } A$. It is claimed that $f(x) = y$. Since x is a critical point of the differentiable function h, we have $(\partial h/\partial x_j)\big|_x = 0$ for all $j = 1, \ldots, m$; i.e.,

$$\sum_{i=1}^{m} 2[y_i - f_i(x)] \frac{\partial f_i}{\partial x_j}\bigg|_x = 0 \qquad \text{for all } j = 1, \ldots, m$$

By i, $Df(x)$ is nonsingular. Thus the only solution of the above system of homogeneous linear equations is the trivial solution. Thus

$$y_i - f_i(x) = 0 \qquad \text{for all } i = 1, \ldots, m$$

which implies $y = f(x)$ as claimed.

$\quad f^{-1} : B \to C$ *is continuous*: This follows from iii.

$\quad f^{-1} : B \to C$ *is of class* C^r : This is carried out inductively. We show that $D(f^{-1})$ is the composition

$$B \xrightarrow{f^{-1}} C \xrightarrow{Df} GL(m) \xrightarrow{\text{matrix inversion}} GL(m)$$

where $GL(m)$ denotes the nonsingular $m \times m$ matrices, which may be topologized as a subspace of $\mathbb{R}^{m^2}$. Now matrix inversion is C^∞ and Df is C^{r-1}. Thus continuity of f^{-1} implies continuity of $D(f^{-1})$, so that f^{-1} is C^1. But then (unless $r = 1$) $D(f^{-1})$ is C^1, so that f^{-1} is C^2. This process continues inductively until we get that $D(f^{-1})$ is C^{r-1}, i.e., f^{-1} is C^r. It stops at this point because Df is assumed only to be C^{r-1}.

Appendix A. The Unproved Theorems 221

To show that $D(f^{-1})$ is the above composition, let $\bar{y} \in B$. By Taylor's theorem,

$$y^* = \bar{y}^* + Df(\bar{x}) \cdot (x - \bar{x})^* + k(x,\bar{x})^* \quad \text{for all } y \in B$$

where $\bar{x} = f^{-1}(\bar{y})$, $x = f^{-1}(y)$, and

$$\lim_{x \to \bar{x}} \frac{|k(x,\bar{x})|}{|x - \bar{x}|} = 0$$

By i, $Df(\bar{x})$ is nonsingular. Let $M = Df(\bar{x})^{-1}$. Then

v. $\quad x^* - \bar{x}^* = M(y - \bar{y})^* - Mk(x,\bar{x})^*$

Let $k_1(y,\bar{y})^* = Mk(f^{-1}(y), f^{-1}(\bar{y}))^*$.

If M is any $m \times m$ matrix then there exists $K \in \mathbb{R}$ such that for each $x \in \mathbb{R}^m$, $|xM^*| \leq K|x|$. To prove this, first notice that matrix multiplication determines a continuous function $S^{m-1} \to \mathbb{R}^m$ defined by $x \mapsto xM^*$. Since S^{m-1} is compact, by Theorems 3.2 and 3.3, there is a number K with $|xM^*| \leq K$ for each $x \in S^{m-1}$. It is claimed that $|xM^*| \leq K|x|$ for each $x \in \mathbb{R}^m$. This is obvious for $x = 0$, and if $x \in \mathbb{R}^m - \{0\}$, then $x/|x| \in S^{m-1}$ so $|(x/|x|)M^*| \leq K$. Since matrix multiplication is linear, the $|x|$ in the denominator may be transferred to the other side.

Returning to our particular matrix M, we have

$$|k_1(y,\bar{y})| = |k(x,\bar{x})M^*|$$
$$\leq K|k(x,\bar{x})|$$

so that

$$\frac{|k_1(y,\bar{y})|}{|y - \bar{y}|} \leq K \frac{|x - \bar{x}|}{|y - \bar{y}|} \frac{|k(x,\bar{x})|}{|x - \bar{x}|}$$

$$\leq 2K \frac{|k(x,\bar{x})|}{|x - \bar{x}|} \quad \text{by iii}$$

Therefore, as $y \to \bar{y}$, $|k_1(y,\bar{y})|/|y-\bar{y}| \to 0$. Thus from v, we have

$$f^{-1}(y)* = f^{-1}(\bar{y})* + M(y - \bar{y})* + k_1(y,\bar{y})*$$

so $D(f^{-1})(\bar{y}) = M = [Df(f^{-1}(y))]^{-1}$, which is the claimed composition. □

THEOREM 9.6. Every compact manifold possesses a Morse function.

Proof: Refer to Ref. 10 or Ref. 14.

In fact, almost every differentiable function with domain a compact manifold and range $\mathbb{R}$ is a Morse function in a sense made precise in Refs. 10 and 14.

The basic idea of the proof is as follows. If $f : \mathbb{R}^m \to \mathbb{R}$ is C^2, then for almost every linear map $\lambda : \mathbb{R}^m \to \mathbb{R}$, the function $f + \lambda : \mathbb{R}^m \to \mathbb{R}$ has only nondegenerate critical points. This enables one to take an arbitrary real-valued differentiable function on a compact manifold and to modify it chart by chart until it becomes a Morse function. Since the manifold is compact, each point has its value changed only finitely many times, since only finitely many charts are needed to cover the manifold. Of course, one needs to be sure that the modification within one chart does not undo what has already been done in a previous chart. □

In addition to the above, the fundamental existence and uniqueness theorem for differential equations was required in the proof of Theorem 10.2. The following is a statement of this result. We treat points of $\mathbb{R}^{m+1}$ as pairs of points, the first from $\mathbb{R}$ and the second from $\mathbb{R}^m$; thus we identify $\mathbb{R}^{m+1}$ with $\mathbb{R} \times \mathbb{R}^m$.

THEOREM. Let $F : C \to \mathbb{R}^m$ be a continuous function, where C is some cube in $\mathbb{R}^{m+1}$ such that for some $L \in \mathbb{R}$,

$$|F(t,x) - F(t,y)| \leq L|x - y| \quad \text{for all } (t,x),(t,y) \in C \quad (*)$$

Let $(t_0, x_0) \in \text{Int } C$. Then there exist $\delta > 0$ and a unique function $\gamma : [t_0 - \delta, t_0 + \delta] \to \mathbb{R}^m$ such that $\gamma(t_0) = x_0$ and

Appendix A. The Unproved Theorems

$$\gamma'(t) = F(t,\gamma(t)) \quad \text{for all } t \in [t_0 - \delta, t_0 + \delta]$$

The condition (*) is known as a Lipschitz condition. In the context of Theorem 10.2, the function F is independent of t, so C is taken to be a cube in $\mathbb{R}^m$. Satisfaction of the Lipschitz condition in this case follows from exercise 15.1.

Variations on the above statement together with its proof may be found in any text covering the theory of differential equations, for example, Ref. 3. Here is a brief outline of the proof.

The number δ depends on L, the size of C, and the size of F(C). One then lets

$$\Gamma = \{\gamma : [t_0 - \delta, t_0 + \delta] \to \mathbb{R}^m \mid \gamma \text{ is continuous}\}$$

and defines a transformation $T : \Gamma \to \Gamma$ by

$$T(\gamma)(t) = x_0 + \int_{t_0}^{t} F(t,\gamma(t))\, dt \quad \text{for all } t \in [t_0 - \delta, t_0 + \delta]$$

A notion of distance is defined on Γ and it is shown using (*) that $T(\gamma_1)$ and $T(\gamma_2)$ are somewhat closer together than γ_1 and γ_2. From this and a compactness property of Γ, it follows that there is a unique $\gamma \in \Gamma$ for which $T(\gamma) = \gamma$. In fact, if we choose any $\gamma_1 \in \Gamma$ and define the sequence (γ_n) inductively by letting $\gamma_n = T(\gamma_{n-1})$, then $\gamma = \lim_{n \to \infty} \gamma_n$. The compactness property of Γ together with the property of T tell us that the limit exists. The property of T implies its independence of the choice of γ_1. Further,

$$T(\gamma) = T(\lim_{n \to \infty} \gamma_n) = \lim_{n \to \infty} T(\gamma_n) = \lim \gamma_{n+1} = \gamma$$

By definition of T, we have

$$\gamma(t) = x_0 + \int_{t_0}^{t} F(t,\gamma(t))\, dt \quad \text{for all } t \in [t_0 - \delta, t_0 + \delta]$$

from which the conclusion immediately follows. □

Lemma 14.5 required the Jordan curve theorem. This theorem states that if $e : S^1 \to S^2$ is an embedding, then $S^2 - e(S^1)$ is disconnected; moreover, $S^2 - e(S^1)$ has exactly two components. One could replace S^2 by $\mathbb{R}^2$ in this statement. A proof of this theorem may be found, for example, in Refs. 8, 11, and 15. In Refs. 8 and 11 it is shown more generally that if $e : S^{n-1} \to S^n$ is a topological embedding then $S^n - e(S^{n-1})$ has exactly two components. This result is known as the Jordan-Brouwer separation theorem.

Lemma 14.5 also required the finiteness of the genus of an orientable surface. Here is one way of verifying this fact though, unfortunately, it requires homology theory. Let S be a sphere with n handles, and let $e_i : S^1 \times (-1,1) \to S$ be an embedding (i = 1, ..., m) such that writing $D_i = e_i(S^1 \times (-1,1))$ and $D = \bigcup_{i=1}^m D_i$, we have $i \neq j \Rightarrow D_i \cap D_j = \phi$ and S - D is connected. We show that $m \leq 2n$. Let $C_i = e_i(S^1 \times \{0\})$ and $C = \bigcup_{i=1}^m C_i$. Then $H_1(D, D-C) \approx \mathbb{Z}^m$, and by excision, $H_1(D, D-C) \approx H_1(S, S-C)$. Thus the portion $H_1(S) \to H_1(S, S-C) \to \tilde{H}_0(S-C)$ of the reduced exact sequence of the pair $(S, S-C)$ reduces to $\mathbb{Z}^{2n} \to \mathbb{Z}^m \to 0$. Exactness implies that $m \leq 2n$.

From time to time, beginning in Chap. 3, we appealed to invariance of domain. This result says that if U is a connected open subset of $\mathbb{R}^n$ and $f : U \to \mathbb{R}^n$ is a continuous injection, then f(U) is open and f is an embedding. As is the case in Refs. 8 and 11, it is usually deduced as a consequence of the Jordan-Brouwer separation theorem, the idea being that one chooses, for each $x \in U$, a small ball in U centered at x. The image under f of the boundary of this ball divides $\mathbb{R}^n$ into exactly two pieces, one of which is the image under f of the interior of the ball. Thus f(U) must be open.

Appendix B

FURTHER TOPICS

In this appendix, we suggest a number of topics which the student might like to pursue in addition to what was covered in the preceding chapters. In each case the content of the work is approximately equivalent to one of the chapters of this text.

FURTHER POINT SET TOPOLOGY

The Hausdorff property is one example of a separation property. The basic separation properties are of the following form. Let X be a topological space; say that the disjoint subsets A and B of X are separated provided they have disjoint neighborhoods. X is Hausdorff provided distinct pairs of points are separated. X is *regular* provided the disjoint subsets A and B are separated whenever A consists of a single point and B is closed. X is *normal* provided the disjoint subsets A and B are separated whenever they are both closed.

Standard topology textbooks consider relationships between the various separation properties together with their connections with other topological properties. An example of the latter, which continues a theme touched on in Theorem 3.6, is the theorem which asserts that compact Hausdorff spaces are normal.

Normal spaces are, in addition to being an important class of topological spaces, very interesting in that they have a number of curious equivalent conditions. For example, Urysohn's lemma asserts that X is normal if and only if for every pair A,B of disjoint closed subsets of X, there is a continuous function $f : X \to [0,1]$ with $f(A) = 0$ and $f(B) = 1$.

Refs. 5, 11, and 17 contain a wealth of information on these and related matters.

CLASSIFICATION OF NONORIENTABLE SURFACES

There are two methods of constructing nonorientable surfaces using the five basic traces of Theorem 13.4. One can use only the four traces used in Chap. 14, but it is necessary to ensure that at least one of the connecting type (1,1) surgeries is performed in the nonorientable manner as noted after Theorem 13.4. Alternatively, one could use at least one twisting surgery of type (1,1). Even if we use all five basic traces, Lemma 14.1 still goes through. An analogue of Lemma 14.2 is the following.

LEMMA. Let $f : S \to \mathbb{R}$ be a Morse function on a nonorientable surface S such that f has only one maximum and one minimum. Then there is another Morse function $g : S \to \mathbb{R}$ with the same critical points as f and the same indices, but such that all the surgeries of type (1,1) determined by g are of the twisting kind.

Once this lemma has been proved, it is a short step to the classification of nonorientable surfaces. They are classified by the number of critical points of index 1, which is the genus of the surface. One can find a circle embedded in the trace of a twisting type (1,1) surgery which does not disconnect the trace.

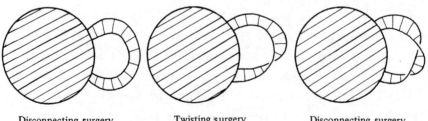

Disconnecting surgery   Twisting surgery   Disconnecting surgery

FIGURE 81

Appendix B. Further Topics   227

To prove the lemma, we introduce another way of viewing a surgery. The trace effect of performing a surgery of type (1,1) on the boundary of a 2-ball is to add a handle. As shown in Fig. 81, if the surgery is disconnecting, then the handle has no twist (or an even number), and if the surgery is twisting, then the handle has one twist (or an odd number).

Of course the boundaries of the first and third pictures in Fig. 81 are a pair of circles, and of the second, a single circle, as expected. To perform several surgeries of type (1,1), we add on several handles, inserting a twist if prescribed, disconnecting one of the bounding circles if prescribed, or connecting two of the bounding circles if prescribed. Fig. 82 illustrates several such surgeries. The numbers within a handle indicate the order in which the handle is added. Handles 1, 3, 4, and 5 correspond to disconnecting surgeries, handle 2 to a connecting surgery, and handle 6 to a twisting surgery. Note that handle 5 could be added before any of the others without affecting the qualities of the surgeries (this corresponds to lowering the critical point as in Lemma 14.2). On the other hand, interchanging the order of surgeries 1 and 2 interchanges their roles as well; compare with the discussion after the proof of Lemma 14.2, where a lowering of the upper saddle below the lower changes the quality of the surgery in exactly the same way.

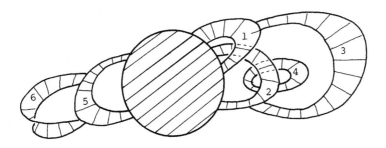

FIGURE 82

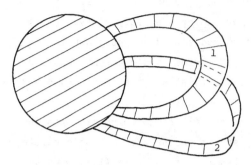

FIGURE 83

We now sketch the proof of the lemma. If a connecting surgery of type (1,1) is performed so that it results in a nonorientable surface as in the last picture of Fig. 64, then not only must the handle connect two circles but also it must have a twist in it. Fig. 83 shows such an effect. Handle 1 corresponds to a disconnecting surgery, and handle 2 to a connecting surgery performed so as to give a nonorientable manifold. If we interchange the surgeries (as in Lemma 14.2: lower the upper critical point) then we will perform a twisting surgery first. The second surgery will also, of necessity, be a twisting surgery since the boundary of the disk plus handle 2 is a single circle.

If the situation of the previous paragraph does not arise then there must be a twisting surgery. If the only surgeries of type (1,1) are twisting surgeries, then there is nothing to prove. Otherwise there must be some disconnecting surgeries and an equal number of connecting surgeries. We show how to change a combination of one of each of these three kinds into a combination of three twisting surgeries. Whatever the order in which these three surgeries is performed, we can rearrange the handles to appear as in the first picture of Fig. 84. The second picture shows a rearrangement of these handles so that, when added in the order indicated, they are all twisted handles. Note that we have also changed the Morse function on part of the original ball; in fact, part of the original ball forms one of the handles in the new arrangement.

Appendix B. Further Topics

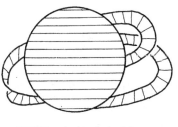 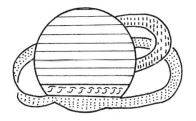

══ original ball   ░░░ first handle   ▓▓▓ second handle
ℐℐℐℐ third handle

FIGURE 84

SARD'S THEOREM

This asserts that the set of critical values of a differentiable function $f : M^m \to N^n$ has n-dimensional Hausdorff measure 0. A proof of this theorem can be found in most standard differential topology texts, for example, Refs. 1 or 10.

Among the many interesting consequences of Sard's theorem, here are two.

APPLICATION 1. Let $f_1, \ldots, f_n : \mathbb{R}^m \to \mathbb{R}$ be n differentiable functions. Then for almost all choices of n real numbers $a_1, \ldots, a_n$, the set

$$\{x \in \mathbb{R}^m \mid \text{for all } i = 1, \ldots, n, \; f_i(x) = a_i\}$$

is an $(m - n)$-submanifold of $\mathbb{R}^m$.

Thus a single (differentiable) equation in $\mathbb{R}^m$ usually determines an $(m - 1)$-manifold; two equations determine an $(m - 2)$-manifold, the intersection of the two $(m - 1)$-manifolds, etc. For example, $x^2 + y^2 = a$ determines a curve in $\mathbb{R}^2$ for all $a \neq 0$ (the empty curve if $a < 0$).

Proof: To verify this first application, as well as to give meaning to the expression "for almost all choices," we apply Sard's theorem to $f : \mathbb{R}^m \to \mathbb{R}^n$ defined by $f(x) = (f_1(x), \ldots, f_n(x))$. Then for almost all $a \in \mathbb{R}^n$, i.e., for all a not in a set of n-dimensional Hausdorff measure 0, a is a regular value of f; thus by Corollary 9.2, $f^{-1}(a)$ is an $(m - n)$-submanifold of $\mathbb{R}^m$. □

APPLICATION 2. Let $f : B^n \to B^n$ be differentiable. Then there exists $x \in B^n$ such that $f(x) = x$.

Proof: This result, known as Brouwer's fixed point theorem, may be proved as follows. Suppose not. Then we may define a differentiable function $g : B^n \to S^{n-1}$ as follows. Let $g(x)$ be that point of S^{n-1} where the line segment from $f(x)$ through x meets S^{n-1} (see Fig. 85). Note that if $x \in S^{n-1}$ then $g(x) = x$. By Sard's theorem, g has at least one (in fact uncountably many!) regular value, say a. Consider $g^{-1}(a)$ (see Fig. 86). By Corollary 9.2, $g^{-1}(a)$ is a 1-submanifold-with-boundary of B^n, the boundary of $g^{-1}(a)$ lying within $\partial B^n = S^{n-1}$. Since g is the identity on S^{n-1}, a is the only point of S^{n-1} lying in $g^{-1}(a)$. Thus the 1-manifold with boundary $g^{-1}(a)$ has only one boundary point. This is impossible since it is a closed, hence compact, subset of B^n, and the only compact 1-manifolds with boundary are combinations of circles and closed intervals, which contain an even number of boundary points. □

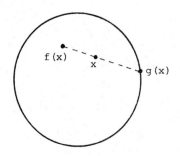

FIGURE 85

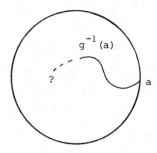

FIGURE 86

Appendix B. Further Topics

Remark In fact, Brouwer's fixed point theorem is more general than the above result. The function f need only be continuous to ensure the existence of a fixed point.

HAUSDORFF DIMENSION

This is a concept of dimension which arises naturally from the concept of Hausdorff measure 0 . First note that when we defined q-dimensional Hausdorff measure 0 in Chap. 15, there was no requirement that q be an integer, the definition being equally valid for noninteger values of q .

If X is any nonempty subset of $\mathbb{R}^n$, then

$$\{q \mid \theta_q(X) = 0\}$$

is bounded below by 0 . The greatest lower bound of this set is called the *Hausdorff dimension* of X . It was observed in Chap. 15 that if q > n then $\theta_q(X) = 0$, so the Hausdorff dimension of X lies between 0 and n . If X is an open subset of $\mathbb{R}^n$ then the Hausdorff dimension of X is n . However most subsets of $\mathbb{R}^n$ have Hausdorff dimension less than n , perhaps even a noninteger.

Examples Any countable subset of $\mathbb{R}^n$ has q-dimensional Hausdorff measure 0 for each q > 0 , so has Hausdorff dimension 0 . Any compact m-manifold has Hausdorff dimension m . There are many weird sets of noninteger Hausdorff dimension.

Cantor's ternary set C is defined as follows. Let $C_0 = [0,1]$, $C_1 = [0,1/3] \cup [2/3,1]$, $C_2 = [0,1/9] \cup [2/9,1/3] \cup [2/3,7/9] \cup [8/9,1]$, etc. Each C_i consists of 2^i closed intervals each of length 3^{-i} ; C_{i+1} is obtained from C_i by cutting out the middle third of each of the intervals making up C_i (see Fig. 87). Then $C = \cap_{i=0}^{\infty} C_i$. It can be shown that the Hausdorff dimension of C is $\log 2/\log 3 \doteq 0.63$.

Let K be the following *Koch curve*: the Hausdorff dimension of K is $\log 4/\log 3 \doteq 1.26$. Define inductively a sequence of increasing polygons P_0, P_1, P_2, ... with the boundary of P_i consisting of 3×4^i line segments each of length 3^{-i} . Of necessity P_0

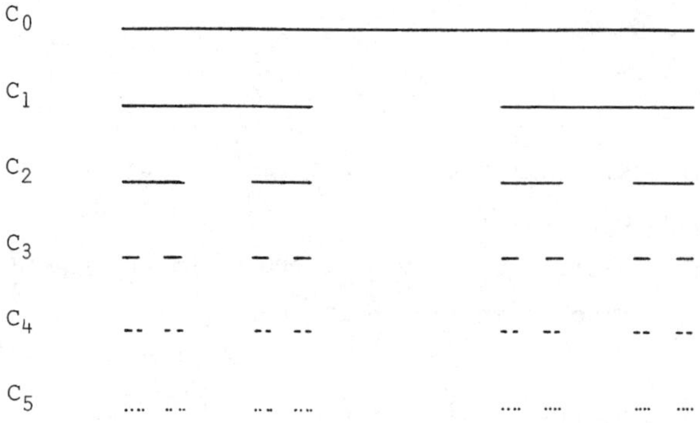

```
c_0  _____    _____

c_1  _____        _____

c_2  _____  _____      _____  _____

c_3  ___ ___  ___ ___    ___ ___  ___ ___

c_4  -- --  -- --   -- --  -- --      -- --  -- --   -- --  -- --

c_5  .... ...  ... ...                  .... ....   .... ....
```

FIGURE 87

is an equilateral triangle with sides of length 1. Given P_i, we divide each side of P_i into thirds and attach to the middle third of each side an equilateral triangle lying outside P_i; P_{i+1} is the union of P_i with these 3×4^i triangles (see Fig. 88). K is defined to be the boundary of $\bigcup_{i=0}^{\infty} P_i$. It is interesting to note that the boundary of P_i has length $3 \times (4/3)^i$, which tends to ∞ as $i \to \infty$. However if we replace 3^{-i}, which is the 1-dimensional measure of a ball of diameter 3^{-i}, by $(3^{-i})^{\log 4/\log 3}$, which is the (log 4/log 3)-dimensional measure of such a ball (up to a constant multiple!) then the (log 4/log 3)-dimensional size of the boundary of P_i is

$$3 \times 4^i \times (3^{-i})^{\log 4/\log 3}$$
$$= 3 \times 4^i \times (3^{\log 4/\log 3})^{-i}$$
$$= 3 \times 4^i \times 4^{-i}$$
$$= 3$$

which is independent of i. This suggests a way of measuring the size of K.

Appendix B. Further Topics

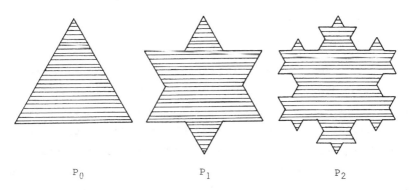

P_0 P_1 P_2

FIGURE 88

Many examples of sets having noninteger Hausdorff dimension, together with relationships with lengths of coastlines (coastlines have dimension about 1.3), river networks, and many other natural phenomena may be found in Ref. 13. Other approaches to the problem of defining the dimension of a set may be found in books on dimension theory, for example, Ref. 12.

DYNAMICAL SYSTEMS

A dynamical system on a manifold is a vector field on the manifold. Theorem 10.2 motivates the second name; by integrating the vector field, we obtain curves, which may be thought of as flow lines, on the manifold. These curves may also be called orbits of the system. As in Theorem 10.2, provided the manifold, say M, is compact, there is a smooth function $\varphi : M \times \mathbb{R} \to M$, called a flow, such that

i. For all $x \in M$, $\varphi \mid \{x\} \times \mathbb{R}$ is the orbit through x.
ii. For all $t \in \mathbb{R}$, $\varphi_t : M \to M$, defined by $\varphi_t(x) = \varphi(x,t)$, is a diffeomorphism.

Say that two vector fields ξ and η on M are equivalent, $\xi \sim \eta$, if there is a homeomorphism of M which takes the ξ orbits onto the η orbits. We can topologize the set of all vector fields Ξ on M by decreeing $\xi \in \Xi$ to be near $A \subset \Xi$ provided that for all $p \in M$, the vector $\xi(p)$ is near the set of vectors

$$\{\eta(p) \mid \eta \in A\}$$

and, given charts about p and $\xi(p)$, the jacobian matrix of ξ with respect to these charts is near the corresponding set of matrices obtained from A. The vector field ξ is structurally stable provided that there is a neighborhood of ξ in Ξ consisting solely of vector fields equivalent to ξ. Obviously the set of structurally stable vector fields forms an open set. It would be nice if it were also dense, i.e., its closure were all of Ξ, for then the nonstable vector fields could be effectively ignored. If dim $M \leq 2$, this is the case, but not so in higher dimensions.

The standard idealized vector fields of physics are usually not structurally stable. If one disturbs them a bit they change drastically. For example, the frictionless simple harmonic oscillator or simple pendulum are not structurally stable. Introduce some friction and the situation changes drastically. The advantage of the idealized vector fields is that the resulting differential equations can be solved in closed form whereas the general ones cannot.

There is an interesting discussion of structural stability in Ref. 20, and a discussion of the two-body problem and related central force problems in Ref. 18.

NOTATION

$\forall$	for every
$\exists$	there exists
$\exists!$	there exists a unique
$\not\exists$	there does not exist
$\in$	is an element of
$\notin$	is not an element of
$\subset$	is a subset of
$\hookrightarrow$	the inclusion function
ϕ	the empty set
$\approx$	is diffeomorphic to

If A and B are subsets of some set X, then A - B denotes $\{x \in X \mid x \in A \text{ but } x \notin B\}$.

If $f : X \longrightarrow Y$ is a function and $A \subset X$ and $B \subset Y$ then $f(A) = \{f(x) \in Y \mid x \in A\}$ and $f^{-1}(B) = x \in X \mid f(x) \in B\}$.

$\mathbb{R}$	the set of real numbers
$\mathbb{Q}$	the set of rational numbers
$\mathbb{Z}$	the set of integers
$\mathbb{R}^n$	$\{(x_1,\ldots,x_n) \mid x_i \in \mathbb{R}\}$

$\mathbb{R}^n$ is considered as a subset of $\mathbb{R}^{n+1}$ by identifying $(x_1,\ldots,x_n) \in \mathbb{R}^n$ with $(x_1,\ldots,x_n,0) \in \mathbb{R}^{n+1}$; thus $\mathbb{R}^m \subset \mathbb{R}^n$ when $m \leq n$.

$0 = (0,\ldots,0) \in \mathbb{R}^n$

$|x| = \sqrt{\sum_{i=1}^{n} x_i^2}$ for $x = (x_1,\ldots,x_n) \in \mathbb{R}^n$

$S^n = \{x \in \mathbb{R}^{n+1} \mid |x| = 1\}$, the n-sphere

$B^n = \{x \in \mathbb{R}^n \mid |x| \leq 1\}$, the n-ball

$rB^n = \{x \in \mathbb{R}^n \mid |x| \leq r\}$

$H^n = \{(x_1,\ldots,x_n) \in \mathbb{R}^n \mid x_n \geq 0\}$

$B(x;r) = \{y \in \mathbb{R}^n \mid |x-y| < r\}$, the open n-ball of radius r centered at x.

If $A \subset \mathbb{R}^m$ and $B \subset \mathbb{R}^n$ then $A \times B = \{(x_1,\ldots,x_{m+n}) \in \mathbb{R}^{m+n} \mid (x_1,\ldots,x_m) \in$ and $(x_{m+1},\ldots,x_{m+n}) \in B\}$. Thus $S^1 \times [0,1]$ is a right circular cylinder in $\mathbb{R}^3$ with axis the z axis and lying between the planes $z = 0$ and $z = 1$.

ν	is near (Chap. 1)
2	the discrete space with two elements (Chap. 1)
Cl	the closure of (Chap. 2)
Int	the interior of (Chap. 2)
Fr	the frontier of (Chap. 2)
T^2	the torus (Chap. 3)
K^2	the Klein bottle (Chap. 3)
$Df(x)$	the jacobian matrix (Chap. 4)
$\Delta f(x)$	the jacobian determinant (Chap. 4)
$Hf(x)$	the hessian (Chap. 4)
C^r	a class of differentiability (Chaps 4 and 5)
P^n	projective n-space (Chap. 6)
$C^r(M,N)$	the set of all C^r maps from M to N (Chap. 8)
TM_p	the tangent space to M at p (Chap. 8)
TM	the tangent manifold (Chap. 10)
$M \cup_h N$	adjunction manifold (Chap. 11)
$\chi(M,e)$	the manifold obtained by performing surgery on M along e (Chap. 11)
$\omega(M,e)$	the trace of a surgery (Chap. 12)
$P_{m,n}$	the patch used to construct the trace of a surgery (Chap. 12)
□	the end of a proof

BIBLIOGRAPHY

BIBLIOGRAPHICAL NOTES

Refs. 7 and 19, although different in aim and scope from this book, are at a similar level and cover more or less the same topics; more precisely, they both use a surgery approach to the classification of surfaces. Ref. 15 gives a completely different approach to the classification of surfaces; it considers only polyhedral surfaces where we considered only smooth surfaces. The classification, however, is exactly the same.

Refs. 1, 10, 14, and 16 cover differential topology at a higher level than we have, and could be consulted for more advanced reading. Our treatment of Morse functions and surgery, however, essentially Chaps. 9 to 12, is basically the same as that of Ref. 14.

Our introduction to topology via nearness spaces is the approach advocated in Refs. 2 and 6, where further details can be found. The more traditional approach, together with a lot more point set topology, may be found in Refs. 5, 11, and 17.

1. Th. Bröcker, *Differentiable Germs and Catastrophes*, London Mathematical Society Lecture Note Series 17, Cambridge University Press, Cambridge, 1975.
2. P. Cameron, J. G. Hocking, and S. A. Naimpally, Nearness—a better approach to continuity, American Mathematical Monthly 81(1974), 739-745.
3. E. A. Coddington, *An Introduction to Ordinary Differential Equations*, Prentice Hall, Englewood Cliffs, New Jersey, 1961.

4. M. M. Cohen, Local homeomorphisms of euclidean space onto arbitrary manifolds, Michigan Mathematical Journal 12(1965), 493-498.
5. J. Dugundji, *Topology*, Allyn and Bacon, Boston, 1966.
6. D. B. Gauld, Nearness—a better approach to topology, Mathematical Chronicle 7(1978), 84-90.
7. A. Gramain, *Topologie des surfaces*, Presses Universitaires de France, Paris, 1971.
8. M. J. Greenberg, *Lectures on Algebraic Topology*, W. A. Benjamin, New York, 1967.
9. D. Hilbert and S. Cohn-Vossen, *Geometry and the Imagination*, Chelsea Publishing Company, New York, 1952.
10. M. Hirsch, *Differential Topology*, Graduate Texts in Mathematics 33, Springer-Verlag, New York, 1976.
11. J. G. Hocking and G. S. Young, *Topology*, Addison Wesley, Reading, Massachusetts, 1961.
12. W. Hurewicz and H. Wallman, *Dimension Theory*, Princeton University Press, Princeton, New Jersey, 1941.
13. B. B. Mandelbrot, *Fractals: Form, Chance and Dimension*, W. H. Freeman, San Francisco, 1977.
14. J. W. Milnor, *Lectures on the h-Cobordism Theorem*, Princeton University Press, Princeton, New Jersey, 1965.
15. E. E. Moise, *Geometric Topology in Dimensions 2 and 3*, Graduate Texts in Mathematics 47, Springer-Verlag, New York, 1977.
16. J. R. Munkres, *Elementary Differential Topology*, Annals of Mathematics Studies 54, revised edition, Princeton University Press, Princeton, New Jersey, 1966.
17. J. R. Munkres, *Topology: A First Course*, Prentice Hall, Englewood Cliffs, New Jersey, 1975.
18. S. Smale, Topology and mechanics 1, Inventiones mathematicae 10(1970), 305-331.
19. A. H. Wallace, *Differential Topology: First Steps*, Benjamin, New York, 1968.
20. E. C. Zeeman, *Lecture Notes on Dynamical Systems*, Aarhus Universitet, 1968.

INDEX

A

Adding a handle, 201
Adjunction manifold, 142

B

Basis for a differential
  structure, 54
  for a topology, 18
Boundary, 131
Bounded, 8
  above, 10
Brouwer's fixed point theorem, 230

C

Cantor's ternary set, 231
Chain rule, 42
Classification of compact
  1-manifolds, 171
  of orientable surfaces, 204
  of nonorientable surfaces, 226
Closed set, 20
Closure of a set, 20
Cocountable space, 5
Cofinite space, 5
  topology, 17
Compact space, 29
Completeness axiom, 10
Component, 10
Components of a tangent vector, 100
Concrete space, 4
  topology, 17
Connected space, 9

Connecting surgery of type (1,1), 174
Continuity characterizations, 21
Continuous, 6
Convex set, 209
Coordinate chart, 32
  transformation, 32
Cover, 29
Critical level, 123
  point, 44, 107
    degenerate, 44, 116
    nondegenerate, 44, 116
    index of a, 48, 118
  value, 107
C^r function, 40
  manifold, 54
Cross-cap, 78
Curve, 101

D

Degenerate critical point, 44, 116
Diffeomorphism, 42, 62
Differentiable function, 39, 60
  manifold, 54
Differential structure, 53
  basis for a, 54
  usual on $\mathbb{R}^m$, 55
  usual on S^m, 56
Disconnected, 9
Disconnecting surgery of type (1,1), 174
Disconnection, 9

Discrete space, 4
  topology, 17
Dynamical system, 233

E

Embedding, C^r, 62
  topological, 23
Euclidean space, 3

F

Flow, 233
Frontier of a set, 20

G

Genus, 203
  finiteness of, 224
Gradientlike vector field, 125

H

Half space, 131
Hausdorff dimension, 231
  measure 0, 209, 211
  space, 28
Heine-Borel theorem, 30, 36, 217
Hessian, 44, 116
Homeomorphic, 7
Homeomorphism, 7

I

Immersion, 62
Index of a critical point, 48, 118
Indiscrete space, 4
Integral curve, 134
Interior of a set, 20
Interval, 10
Invariance of domain, 33, 224
Inverse function theorem, 42, 219
Isolated point, 118

J

Jacobian determinant, 42
  matrix, 41
Jordan curve theorem, 224

K

Klein bottle, 34
Koch curve, 231

L

Least upper bound, 10
Level, 123
Lower characteristic embedding, 165
  sphere, 165

M

Manifold with boundary, 131
  differentiable, 54
  tangent, 124
  topological, 32
Morse function, 119, 131
Morse's theorem, 117

N

Near, 3
Nearness relation, 4
  space, 4
Neighborhood, 15
Nondegenerate critical point, 44, 116
Nonorientable surgery of type (1,2), 182
Normal space, 225

O

Open cover, 29
  set, 15, 17
Orbit, 233
Orientability of T^2, 67
  of S^m, 68

Index

Orientable manifold, 67, 72
  surgery of type (1,2), 182
Orientation for a manifold, 67
  preserving, 66, 67
  reversing, 66, 67
Oriented manifold, 67

P

Product of two manifolds, 91, 131
  topology, 24
Projective space, 73
Pythagorean distance, 3

R

Rank of a function at a point, 42
  62,
Regular level, 123
  point, 107
  space, 225
  value, 107

S

Sard's theorem, 229
Separation property, 225
Smooth function, 40
Sphere with handles, 201
Stereographic projection, 33
Structural stability, 234
Subcover, 29
Submanifold, 83
Submersion, 107
Subspace, 6
Surface, 169
Surgery, 148
Surgical descendant, 191

T

Tangent hyperplane, 106
  manifold, 124
  space, 95
  vector, 94, 100
    components of a, 100
Topological invariant/property, 8
  manifold, 32
  space, 17
Topology, 17
  basis for a, 18
Torus, 34
Trace of a surgery, 160
Twisting surgery of type (1,1), 175

U

Upper bound, 10
  characteristic embedding, 165
  sphere, 165
Urysohn's lemma, 225
Usual differential structure on $\mathbb{R}^m$, 55
  on S^m, 56
Usual topology on $\mathbb{R}^n$, 17

V

Vector field, 124
  gradientlike, 125
Velocity vector, 102

W

Whitney's embedding theorem, 215

Mathematics

FUNCTIONAL ANALYSIS (Second Corrected Edition), George Bachman and Lawrence Narici. Excellent treatment of subject geared toward students with background in linear algebra, advanced calculus, physics and engineering. Text covers introduction to inner-product spaces, normed, metric spaces, and topological spaces; complete orthonormal sets, the Hahn-Banach Theorem and its consequences, and many other related subjects. 1966 ed. 544pp. 6⅛ x 9¼. 0-486-40251-7

ASYMPTOTIC EXPANSIONS OF INTEGRALS, Norman Bleistein & Richard A. Handelsman. Best introduction to important field with applications in a variety of scientific disciplines. New preface. Problems. Diagrams. Tables. Bibliography. Index. 448pp. 5⅜ x 8½. 0-486-65082-0

VECTOR AND TENSOR ANALYSIS WITH APPLICATIONS, A. I. Borisenko and I. E. Tarapov. Concise introduction. Worked-out problems, solutions, exercises. 257pp. 5⅝ x 8¼. 0-486-63833-2

AN INTRODUCTION TO ORDINARY DIFFERENTIAL EQUATIONS, Earl A. Coddington. A thorough and systematic first course in elementary differential equations for undergraduates in mathematics and science, with many exercises and problems (with answers). Index. 304pp. 5⅜ x 8½. 0-486-65942-9

FOURIER SERIES AND ORTHOGONAL FUNCTIONS, Harry F. Davis. An incisive text combining theory and practical example to introduce Fourier series, orthogonal functions and applications of the Fourier method to boundary-value problems. 570 exercises. Answers and notes. 416pp. 5⅜ x 8½. 0-486-65973-9

COMPUTABILITY AND UNSOLVABILITY, Martin Davis. Classic graduate-level introduction to theory of computability, usually referred to as theory of recurrent functions. New preface and appendix. 288pp. 5⅜ x 8½. 0-486-61471-9

ASYMPTOTIC METHODS IN ANALYSIS, N. G. de Bruijn. An inexpensive, comprehensive guide to asymptotic methods–the pioneering work that teaches by explaining worked examples in detail. Index. 224pp. 5⅜ x 8½ 0-486-64221-6

APPLIED COMPLEX VARIABLES, John W. Dettman. Step-by-step coverage of fundamentals of analytic function theory–plus lucid exposition of five important applications: Potential Theory; Ordinary Differential Equations; Fourier Transforms; Laplace Transforms; Asymptotic Expansions. 66 figures. Exercises at chapter ends. 512pp. 5⅜ x 8½. 0-486-64670-X

INTRODUCTION TO LINEAR ALGEBRA AND DIFFERENTIAL EQUATIONS, John W. Dettman. Excellent text covers complex numbers, determinants, orthonormal bases, Laplace transforms, much more. Exercises with solutions. Undergraduate level. 416pp. 5⅜ x 8½. 0-486-65191-6

RIEMANN'S ZETA FUNCTION, H. M. Edwards. Superb, high-level study of landmark 1859 publication entitled "On the Number of Primes Less Than a Given Magnitude" traces developments in mathematical theory that it inspired. xiv+315pp. 5⅜ x 8½. 0-486-41740-9

CATALOG OF DOVER BOOKS

CALCULUS OF VARIATIONS WITH APPLICATIONS, George M. Ewing. Applications-oriented introduction to variational theory develops insight and promotes understanding of specialized books, research papers. Suitable for advanced undergraduate/graduate students as primary, supplementary text. 352pp. 5⅜ x 8½.
0-486-64856-7

COMPLEX VARIABLES, Francis J. Flanigan. Unusual approach, delaying complex algebra till harmonic functions have been analyzed from real variable viewpoint. Includes problems with answers. 364pp. 5⅜ x 8½. 0-486-61388-7

AN INTRODUCTION TO THE CALCULUS OF VARIATIONS, Charles Fox. Graduate-level text covers variations of an integral, isoperimetrical problems, least action, special relativity, approximations, more. References. 279pp. 5⅜ x 8½.
0-486-65499-0

COUNTEREXAMPLES IN ANALYSIS, Bernard R. Gelbaum and John M. H. Olmsted. These counterexamples deal mostly with the part of analysis known as "real variables." The first half covers the real number system, and the second half encompasses higher dimensions. 1962 edition. xxiv+198pp. 5⅜ x 8½. 0-486-42875-3

CATASTROPHE THEORY FOR SCIENTISTS AND ENGINEERS, Robert Gilmore. Advanced-level treatment describes mathematics of theory grounded in the work of Poincaré, R. Thom, other mathematicians. Also important applications to problems in mathematics, physics, chemistry and engineering. 1981 edition. References. 28 tables. 397 black-and-white illustrations. xvii + 666pp. 6⅛ x 9¼.
0-486-67539-4

INTRODUCTION TO DIFFERENCE EQUATIONS, Samuel Goldberg. Exceptionally clear exposition of important discipline with applications to sociology, psychology, economics. Many illustrative examples; over 250 problems. 260pp. 5⅜ x 8½.
0-486-65084-7

NUMERICAL METHODS FOR SCIENTISTS AND ENGINEERS, Richard Hamming. Classic text stresses frequency approach in coverage of algorithms, polynomial approximation, Fourier approximation, exponential approximation, other topics. Revised and enlarged 2nd edition. 721pp. 5⅜ x 8½. 0-486-65241-6

INTRODUCTION TO NUMERICAL ANALYSIS (2nd Edition), F. B. Hildebrand. Classic, fundamental treatment covers computation, approximation, interpolation, numerical differentiation and integration, other topics. 150 new problems. 669pp. 5⅜ x 8½. 0-486-65363-3

THREE PEARLS OF NUMBER THEORY, A. Y. Khinchin. Three compelling puzzles require proof of a basic law governing the world of numbers. Challenges concern van der Waerden's theorem, the Landau-Schnirelmann hypothesis and Mann's theorem, and a solution to Waring's problem. Solutions included. 64pp. 5⅜ x 8½.
0-486-40026-3

THE PHILOSOPHY OF MATHEMATICS: AN INTRODUCTORY ESSAY, Stephan Körner. Surveys the views of Plato, Aristotle, Leibniz & Kant concerning propositions and theories of applied and pure mathematics. Introduction. Two appendices. Index. 198pp. 5⅜ x 8½. 0-486-25048-2

CATALOG OF DOVER BOOKS

INTRODUCTORY REAL ANALYSIS, A.N. Kolmogorov, S. V. Fomin. Translated by Richard A. Silverman. Self-contained, evenly paced introduction to real and functional analysis. Some 350 problems. 403pp. 5⅜ x 8½. 0-486-61226-0

APPLIED ANALYSIS, Cornelius Lanczos. Classic work on analysis and design of finite processes for approximating solution of analytical problems. Algebraic equations, matrices, harmonic analysis, quadrature methods, much more. 559pp. 5⅜ x 8½. 0-486-65656-X

AN INTRODUCTION TO ALGEBRAIC STRUCTURES, Joseph Landin. Superb self-contained text covers "abstract algebra": sets and numbers, theory of groups, theory of rings, much more. Numerous well-chosen examples, exercises. 247pp. 5⅜ x 8½. 0-486-65940-2

QUALITATIVE THEORY OF DIFFERENTIAL EQUATIONS, V. V. Nemytskii and V.V. Stepanov. Classic graduate-level text by two prominent Soviet mathematicians covers classical differential equations as well as topological dynamics and ergodic theory. Bibliographies. 523pp. 5⅜ x 8½. 0-486-65954-2

THEORY OF MATRICES, Sam Perlis. Outstanding text covering rank, nonsingularity and inverses in connection with the development of canonical matrices under the relation of equivalence, and without the intervention of determinants. Includes exercises. 237pp. 5⅜ x 8½. 0-486-66810-X

INTRODUCTION TO ANALYSIS, Maxwell Rosenlicht. Unusually clear, accessible coverage of set theory, real number system, metric spaces, continuous functions, Riemann integration, multiple integrals, more. Wide range of problems. Undergraduate level. Bibliography. 254pp. 5⅜ x 8½. 0-486-65038-3

MODERN NONLINEAR EQUATIONS, Thomas L. Saaty. Emphasizes practical solution of problems; covers seven types of equations. ". . . a welcome contribution to the existing literature...."–*Math Reviews*. 490pp. 5⅜ x 8½. 0-486-64232-1

MATRICES AND LINEAR ALGEBRA, Hans Schneider and George Phillip Barker. Basic textbook covers theory of matrices and its applications to systems of linear equations and related topics such as determinants, eigenvalues and differential equations. Numerous exercises. 432pp. 5⅜ x 8½. 0-486-66014-1

LINEAR ALGEBRA, Georgi E. Shilov. Determinants, linear spaces, matrix algebras, similar topics. For advanced undergraduates, graduates. Silverman translation. 387pp. 5⅜ x 8½. 0-486-63518-X

ELEMENTS OF REAL ANALYSIS, David A. Sprecher. Classic text covers fundamental concepts, real number system, point sets, functions of a real variable, Fourier series, much more. Over 500 exercises. 352pp. 5⅜ x 8½. 0-486-65385-4

SET THEORY AND LOGIC, Robert R. Stoll. Lucid introduction to unified theory of mathematical concepts. Set theory and logic seen as tools for conceptual understanding of real number system. 496pp. 5⅜ x 8¼. 0-486-63829-4

CATALOG OF DOVER BOOKS

TENSOR CALCULUS, J.L. Synge and A. Schild. Widely used introductory text covers spaces and tensors, basic operations in Riemannian space, non-Riemannian spaces, etc. 324pp. 5⅜ x 8¼. 0-486-63612-7

ORDINARY DIFFERENTIAL EQUATIONS, Morris Tenenbaum and Harry Pollard. Exhaustive survey of ordinary differential equations for undergraduates in mathematics, engineering, science. Thorough analysis of theorems. Diagrams. Bibliography. Index. 818pp. 5⅜ x 8½. 0-486-64940-7

INTEGRAL EQUATIONS, F. G. Tricomi. Authoritative, well-written treatment of extremely useful mathematical tool with wide applications. Volterra Equations, Fredholm Equations, much more. Advanced undergraduate to graduate level. Exercises. Bibliography. 238pp. 5⅜ x 8½. 0-486-64828-1

FOURIER SERIES, Georgi P. Tolstov. Translated by Richard A. Silverman. A valuable addition to the literature on the subject, moving clearly from subject to subject and theorem to theorem. 107 problems, answers. 336pp. 5⅜ x 8½. 0-486-63317-9

INTRODUCTION TO MATHEMATICAL THINKING, Friedrich Waismann. Examinations of arithmetic, geometry, and theory of integers; rational and natural numbers; complete induction; limit and point of accumulation; remarkable curves; complex and hypercomplex numbers, more. 1959 ed. 27 figures. xii+260pp. 5⅜ x 8½. 0-486-63317-9

POPULAR LECTURES ON MATHEMATICAL LOGIC, Hao Wang. Noted logician's lucid treatment of historical developments, set theory, model theory, recursion theory and constructivism, proof theory, more. 3 appendixes. Bibliography. 1981 edition. ix + 283pp. 5⅜ x 8½. 0-486-67632-3

CALCULUS OF VARIATIONS, Robert Weinstock. Basic introduction covering isoperimetric problems, theory of elasticity, quantum mechanics, electrostatics, etc. Exercises throughout. 326pp. 5⅜ x 8½. 0-486-63069-2

THE CONTINUUM: A CRITICAL EXAMINATION OF THE FOUNDATION OF ANALYSIS, Hermann Weyl. Classic of 20th-century foundational research deals with the conceptual problem posed by the continuum. 156pp. 5⅜ x 8½. 0-486-67982-9

CHALLENGING MATHEMATICAL PROBLEMS WITH ELEMENTARY SOLUTIONS, A. M. Yaglom and I. M. Yaglom. Over 170 challenging problems on probability theory, combinatorial analysis, points and lines, topology, convex polygons, many other topics. Solutions. Total of 445pp. 5⅜ x 8½. Two-vol. set. Vol. I: 0-486-65536-9 Vol. II: 0-486-65537-7

Paperbound unless otherwise indicated. Available at your book dealer, online at www.doverpublications.com, or by writing to Dept. GI, Dover Publications, Inc., 31 East 2nd Street, Mineola, NY 11501. For current price information or for free catalogues (please indicate field of interest), write to Dover Publications or log on to www.doverpublications.com and see every Dover book in print. Dover publishes more than 500 books each year on science, elementary and advanced mathematics, biology, music, art, literary history, social sciences, and other areas.